Study Guide

Rachael Henriques Porter
Le Moyne College

Foundations of College Chemistry

Fourteenth Edition
Alternate Fourteenth Edition

Morris Hein
Mount San Antonio College

Susan Arena
University of Illinois Urbana–Champaign

WILEY

For my boys, K, K, C, C
       –R.H.P.

Content Editor: Alyson Rentrop
Senior Production Editor: Elizabeth Swain

This book was typeset by Thomson Digital. The cover and text were printed by Bind-Rite.

To order books or for customer service, please call 1-800-CALL-WILEY (225-5945).

ISBN 978-1-118-28900-6

Printed in the United States of America

10 9 8 7 6 5 4 3 2 1

Preface

This study guide has been prepared to accompany *Foundations of College Chemistry, Fourteenth Edition* and *Foundations of College Chemistry, Alternate Fourteenth Edition,* both by Hein and Arena. The guide selects certain key concepts from the text, and provides the student with a means of self-evaluation for determining how well he or she understands the material of each chapter.

By presenting a slightly different viewpoint and emphasis, the study guide provides students with an approach other than that of the textbook toward mastery of the subject matter. Because this guide is an auxiliary student-oriented aid to complement the textbook, no new material is presented.

The chapters in the guide arc organized as follows: The chapter headings are the same as in the text and are followed by a section called "Selected Concepts Revisited" that recaps some of concepts that many students in the past have had difficulty grasping. A "Common Pitfalls to Avoid" section is intended to make students aware of common mistakes previously made by students. By reading this section before completing the self-evaluation section hopefully students will avoid making similar mistakes. A brief recap or summary of the chapter content then preceeds the self-evaluation section. The self-evaluation section provides various exercises that allow the student to check his or her understanding of the text chapter material. Students are asked to complete fill-in responses or to choose an appropriate response. Tables to be constructed and problems to be solved are presented. Nomenclature, equation balancing, and use of the periodic table are covered. Most chapters have challenge problems at the end of the self-evaluation section. These will usually be a little more difficult or complex than the other exercises in the chapter. Additionally, the explanations about the answers will be briefer and will assume that the student has a good grasp of the basic material covered up to that particular point. Complete answers to questions and solutions to problems are given at the end of each chapter, to provide immediate reinforcement or correction of errors.

In addition to the usual exercises, several puzzles are included in the guide to give the student a break and to check his or her vocabulary skills. The student will probably need a periodic table to help in solving the puzzles.

The author visualizes using the study guide as follows. The student would (1) read the assigned text chapter; (2) go through the associated chapter in the study guide including the self-evaluation process; (3) return to the assigned homework problems, lab experiments, and other instructor-generated activities; and (4) be tested on predetermined performance objectives for each chapter.

To the Student

This guide, which accompanies the textbook *Foundations of College Chemistry, Thirteenth Edition* and *Foundations of College Chemistry, Alternate Thirteenth Edition*, both by Hein and Arena, is a student-oriented self-study guide. It has been prepared to help you evaluate your grasp of certain concepts in the textbook; you can then identify trouble spots and ask your instructor for individual help.

What is the place of the study guide in the total chemistry program? Your instructor will hand out reading assignments from the textbook, which will be followed by homework questions and problems and laboratory experiments. After classroom discussion and problem sessions, you will be tested on the material. So, when do you use the study guide? A typical student might first read the assigned chapter in the text (with rereading, if necessary), and then turn to the study guide to check for common misconceptions or easily avoidable mistakes. Completing the self-evaluation section in the study guide then serves as a check for your understanding of the key concepts. All of the answers and solutions are given so that any errors can be quickly spotted. If a particular concept or type of problem doesn't make sense, the student can return to the textbook. Then the student is ready to work the homework assignment, knowing that he or she understands the basic concepts in the chapter. The week's lab experiment is performed, followed by the chapter quiz.

Each chapter of the study guide has questions to answer and problems to solve and is organized as follows: The "Selected Concepts Revisited" section provides students with the opportunity to review some (not all) of the concepts presented in the chapter that have traditionally confused students or are more emphasized, and provides some helpful hints for dealing with the material. The next section "Common Pitfalls to Avoid" highlights some of the mistakes students typically make in answering questions related to this chapter's topics. By making you aware of these possible errors, it is hoped that you will be less likely to repeat these mistakes. The "Self-Evaluation Section" provides various types of exercises that enable you to check your understanding of the key objectives. The answers and solutions to the problems are given at the end of the chapter so that you can find out where your difficulty may have occurred. The challenge problems at the end of each chapter are usually a bit more difficult or complex than the rest of the self-evaluation exercises. You should work on these after you feel comfortable with the other material. The answers to the challenge problems may not be as thorough as the other answers since you're expected to have a good grasp of the basic material before attempting the challenge problems. The last section gives a brief recap of the chapter and indicates where you are going next in the course.

The study guide allows you to work on your own and make efficient use of your time. It should assist you in finding out whether you have learned the basic concepts.

Best wishes for success in your study of chemistry.

Contents

CHAPTER ONE

An Introduction to Chemistry

RECAP SECTION

The objectives of Chapter 1 are to introduce you to the broad field of science known as chemistry. Use this opportunity to familiarize yourself with the terms and definitions mentioned, learning the language of chemistry. Start picturing nature at the microscopic level as it will enable you to better understand chemical behavior.

By the end of this chapter, you should be able to describe the steps involved in the scientific method, determine the states of matter and identify their characteristics, know the differences between pure substances and the types of mixtures, and recognize the mathematical concepts, mostly arithmetic and algebraic, that you need to brush up on.

SELF-EVALUATION SECTION

We will begin your use of the study guide with some straight-forward exercises followed by a quick mathematics section. As mentioned in the introduction, the use of mathematics is very important for your success in chemistry.

1.   The scientific method is a general approach to answer questions or solve problems. Each of the following statements is missing an "action verb" which is to be selected from the list below.

Use these verbs to fill in the blanks:

> formulate
> modify
> analyze
> plan and do
> collect

(1)   _____ the facts or data that are relevant to the question which is usually done by experimentation.

(2)   _____ the data to find trends.

(3)   _____ a hypothesis that will account for the accumulated data and can be tested by further experimentation.

(4)   _____ additional experiments to test the hypothesis.

(5)   _____ the hypothesis as necessary so that it is compatible with all pertinent data.

2.   Indicate whether each statement is an observation, a hypothesis, an experiment, or a theory. Note that these statements are in a random order.

(1)   _____ The cold soda made a fountain 12 inches high while the hot soda made a fountain 29 inches high.

(2)   _____ Using a cold bottle of soda will cause the height of the fountain to be lower than that of a hot bottle of soda.

(3) ___observation___ There was no measurable difference in the height of the soda fountains of the two bottles when different amounts of candy were used.

(4) ___hypothesis___ Using more candy will make the height of the soda fountain increase.

(5) ___experiment___ Take two similar bottles of soda and place one in the fridge and the other in the sun. After 1 hour, put the bottles side-by-side with a meter stick attached to each bottle. Then open both soda bottles and drop the same number of chewable mint candy pieces into each bottle. Measure the height of the fountain that erupts from the bottle.

(6) ___hypothesis___ Diet soda will cause the height of the fountain to be lower than that of a regular soda, all other conditions being equal.

(7) ___theory___ Chewable mint candy pieces dropped into a just opened bottle of carbonated soda will cause a fountain of soda to erupt from the bottle. The height of the fountain may be affected by the type of carbonated soda used, temperature, and the number of candy pieces used.

(8) ___experiment___ Two identical bottles of soda were set up. 10 candy pieces were dropped into one bottle and 20 candy pieces were dropped into the second bottle.

3. Rearrange the statements in question 2 to demonstrate the scientific method. What would be the next logical step?

4. Match the correct definition to each of the following terms.

(1) macroscopic   _____ molecular view of substances

(2) amorphous   __6__ indefinite shape, definite volume

(3) mixture   __1__ the big picture; everyday objects

(4) gas   _____ containing two or more physically distinct phases

(5) microscopic   _____ a solid lacking a regular internal geometric pattern

(6) liquid   _____ indefinite shape, no fixed volume

(7) heterogeneous   _____ contains two or more substances

5. Fill in the blank.

(1) If a solid is not amorphous, it is _____.

(2) If a mixture is not heterogeneous, it is _____.

(3) If a material is not a mixture, it is a _____.

6. Examine the following list of items. Some are mixtures, others are elements or compounds. Place an *S* next to the pure substance and an *M* next to the mixtures.

(1) oxygen   ___M___   (2) milk   ___S___

(3) chlorine   ___M___   (4) air   ___S___

(5) sodium chloride   ___M___   (6) soil   ___S___

(7) wood   ___M___   (8) rubber   ___S___

(9) gasoline   ___M___   (10) water   ___M___

7. All matter in the universe can be classified into one of three states – gas, liquid, or solid. Determine to which of the three states of matter each of the following descriptions relates.

	Description	States of Matter
(1)	Has a definite fixed shape.	_____
(2)	Particles flow over each other while retaining fixed volume.	_____
(3)	Exerts a pressure on all walls of the container.	_____
(4)	Exhibits no or very slight compressibility.	_____
(5)	Particles move independently of each other.	_____
(6)	Particles arranged in regular, fixed geometric pattern.	_____
(7)	Exhibits slight compressibility.	_____
(8)	Exhibits very high compressibility.	_____

8. As a good way to check your readiness to begin a course in chemistry, work through the following mathematical problems. Many students have found it helpful in determining areas to be reviewed.

Circle the response for each of the questions.

   (a) The problem is worked correctly.

   (b) The problem has been worked incorrectly.

(1)	$4/6 = 0.33$	a b	(2)	$1/100 = 0.001$	a b		
(3)	$1/2 \div 1/4 = 2$	a b	(4)	$25 : 3.9 = 6.4 : 1$	a b		
(5)	$(5 \times 10^{-3})(3 \times 10^{5}) = 1.5 \times 10^{2}$	a b	(6)	If $\dfrac{1.5}{x} = \dfrac{3.0}{6.4}$, then $x = 0.31$	a b		
(7)	$(2 \times 10^{3})(5 \times 10^{2}) = 10^{6}$	a b	(8)	$(g/mol)/mol = g$	a b		
(9)	$5389/1000 = 5389000$	a b	(10)	$\dfrac{1}{z/y}$ is the same as $\dfrac{y}{z}$	a b		
(11)	$1/15 = 0.067$	a b	(12)	$20/5 = 0.40$	a b		
(13)	$1/3 - 1/4 = 1/12$	a b	(14)	$6^{2} \times 6^{3} = 6^{6}$	a b		
(15)	If $\dfrac{x}{2.5} = \dfrac{3}{7.5}$, then $x = 1$	a b	(16)	The reciprocal of $1/2 = 2$	a b		
(17)	$\dfrac{a^{7}}{a^{10}} = a^{-3}$	a b	(18)	$0.0054 = 5.4 \times 10^{3}$	a b		
(19)	$mol/L \times L = mol$	a b	(20)	If $\dfrac{P_1 V_1}{T_1} = \dfrac{P_2 V_2}{T_2}$; then $T_2 = \dfrac{P_2 V_2}{T_1 P_1 V_1}$	a b		

9. Calculate or provide the answer for each of the following problems.

   (1) $\dfrac{2 \times 4 \times 6}{3 \times 5} = ?$

   (2) The cube root of $8 = ?$

   (3) 1594.81 rounded to four figures $= ?$

   (4) If $PV = nRT$, and $n = m/MM$, then combine the equations to get a new equation, $MM = ?$

(5) If it takes 1 calorie to heat 1 gram of water 1°C, how many calories will be required to heat 3 grams of water by 5°C?

(6) Let $T_1 \times N_1 = T_2 \times N_2$; If $T_1 = 5, N_1 = 3$, and $T_2 = 6$, then $N_2 = $?

(7) $\sqrt{25} = $?

(8) $5^4/5^3 = $?

(9) rearrange $PV = nRT$ to be in the form of $R = $?

10. Indicate if the following statements are true or false; some of these equations you will encounter later on in the textbook. You do not need to know the equations to be able to answer the questions. These questions are focusing on your ability to understand directly or inversely proportional relationships.

(1) $c = \lambda v$; when c is kept constant, if λ increases, then v should increase also.

(2) $E = h v$, where h is a constant. E is therefore indirectly proportional to v.

(3) $E = hc/\lambda$, where h and c are constants. If λ increases, E will decrease.

(4) Concentration = mol/vol. Concentration is therefore directly proportional to volume.

(5) Concentration = mol/vol. When concentration is kept constant, moles are directly proportional to volume.

(6) Density = mass/volume. If two items have the same density, then the heavier item will take up less space.

(7) $PV = nRT$; n and T are indirectly proportional to each other.

(8) $PV = nRT$; P and n are indirectly proportional to each other.

(9) $MM = dRT/P$; d and P are indirectly proportional to each other.

(10) Density = mass/volume; If two items have the same mass, the more dense item will have a smaller volume.

Check your answers for the previous three questions. Students who miss more than half of the problems will likely have a difficult time with the mathematically related material in the text without an intensive review and extra work outside of class. Keep this in mind if you found that many of the problems were unfamiliar to you or you responded incorrectly. If you missed more than a quarter of the problems, you should study thoroughly the Mathematical Review Appendix at the back of the text.

The importance of problem solving to the science of chemistry cannot be over-emphasized. You will be working problems within the text material, solving homework problems, and using this study guide to evaluate your ability to solve certain specific types of problems. Many of the laboratory experiments also involve problem solving. The topics of scientific notation, significant figures and rounding-off numbers are covered thoroughly in Chapter 2 of the text. Remember that additional helpful information which will be of use throughout your study of chemistry is found in Appendix I of the text.

ANSWERS TO QUESTIONS

1.   (1) collect   (2) analyze   (3) formulate   (4) plan and do   (5) modify

2.   (1) observation   (2) hypothesis   (3) observation   (4) hypothesis
     (5) experiment   (6) hypothesis   (7) theory   (8) experiment

3. Two answers are possible: (7), (2), (5), (1), (4), (8), (3), (6) or (7), (4), (8), (3), (2), (5), (1), (6). The next logical step would be to design an experiment to test diet soda versus regular soda, all other conditions being equal.

4. (5), (6), (1), (7), (2), (4), (3)

5. (1) crystalline   (2) homogeneous   (3) pure substance

6. (1) M   (2) S   (3) M   (4) S
    (5) M   (6) S   (7) M   (8) S
    (9) M   (10) M

7. (1) solid   (2) liquid   (3) gas   (4) solid
    (5) gas   (6) solid   (7) liquid   (8) gas

8. (1) b   (2) a   (3) a   (4) a
    (5) b   (6) b   (7) a   (8) b
    (9) b   (10) a   (11) a   (12) b
    (13) a   (14) b   (15) a   (16) a
    (17) a   (18) b   (19) a   (20) b

9. (1) 3.2   (2) 2   (3) 1595   (4) $MM = mRT/PV$
    (5) 15 calories   (6) 2.5   (7) 5   (8) 5
    (9) $R = PV/nT$

10. (1) F   (2) F   (3) T   (4) F
    (5) T   (6) F   (7) T   (8) F
    (9) F   (10) T

    (1) $E = h\nu$, where h is a constant. E is therefore indirectly proportional to ν.

    (2) $E = hc/\lambda$, where h and c are constants. If λ increases, E will decrease.

    (3)   Concentration = mol/vol. Concentration is therefore directly proportional to volume.

    (4)   Concentration = mol/vol. When concentration is kept constant, moles are directly proportional to volume.

    (5)   Density = mass/volume. If two items have the same density, then the heavier item will take up less space.

    (6) $PV = nRT$; n and T are indirectly proportional to each other.

    (7) $PV = nRT$; P and n are indirectly proportional to each other.

    (8) $MM = dRT/P$; d and P are indirectly proportional to each other.

    (9)   Density = mass/volume; If two items have the same mass, the more dense item will have a smaller volume.

CHAPTER TWO

Standards for Measurement

SELECTED CONCEPTS REVISITED

Numbers expressed in scientific notation have two parts (use 3.4×10^{-5} as an example)
- the part showing the significant figures (generally only one digit before the decimal). (3.4 indicates two significant figures)
- the part showing the location of the decimal point (the relative magnitude of the number). A negative power of 10 means the original number is less than 1. ($\times 10^{-5}$ indicates the decimal point was moved five spaces)

An important issue involved with all these measurements is significant figures. Bear in mind the information found in significant figures. Significant figures essentially indicate the precision to which an instrument was read or a calculation was based on. The last digit of a measurement has some degree of uncertainty and, in direct measurements, is often estimated. Exact numbers have no degree of uncertainty (that is, they are known exactly) and therefore have infinite significant figures.

Most often significant figures will be limited by numbers read in a measurement. Zeros in significant figures can be summarized by looking at these two examples; numbers in bold are significant.

     0.00**45060**   **301**000

Leading zeros are NOT significant; middle zeros are ALWAYS significant; *Trailing zeros are significant ONLY when a decimal point is present.*

Counted integers and defined numbers are exact numbers (no uncertainty).

Quantitative is related to a quantity; a measurement or an amount.

Qualitative is related to description.

You will find it invaluable to carry out some actual measurements to better visualize the relative sizes of metric versus standard units. Some relationships that you may find helpful include the following. The 1500 m race is often called the metric mile (note that 1609 m is actually a more precise conversion). If you buy gas for your car in Canada you can multiply the listed price per liter by four to find the approximate price per gallon (a more precise conversion would involve multiplying by 3.78).

Two terms introduced in Chapter 2 that are used in everyday conversation are *heat* and *temperature*. There is some confusion about the proper usage *of temperature* and *thermal energy*. It is usual to state that the temperature of an object indicates how much heat the object contains. But this is an incorrect notion. Think of two objects of different size but at the same temperature. Which object contains more thermal energy? The larger object, of course. Thus, temperature is only an indication of the degree of "hotness" or "coldness" of an object. The temperature scales are relative, arbitrary scales established with certain reference points. The Celsius scale, for instance, has 0 degrees set at the freezing point of water and 100 degrees set at the boiling point of water.

When converting between units of temperature, consider whether you need to correct only for the difference in the zero point on the temperature scale or if the size of the degree also needs correcting. Converting between Celsius and Kelvin requires a zero point correction; a simple addition or subtraction of the difference in the zero point (273.15). When converting between Fahrenheit and Celsius, both zero point and degree size corrections are needed. An addition or subtraction term (of 32) takes care of the zero point correction, and a multiplication or division term (by 1.8) takes care of the size of the degree correction. Significant figures for temperature values are generally kept at the same precision after unit conversions. For example, if the temperature was measured to the nearest 0.1 °F, after conversion to °C, the temperature is still reported to the nearest 0.1 degree.

The last term defined in this chapter is *density*. Again, we can use either an algebraic form of definition or the English equivalent. The density of a substance is related to its mass and volume. Since all matter in the universe has mass and occupies space or volume, the quantity defined as density is a useful physical characteristic used in describing various substances. The defining equation for density says that the density of a substance is equal to the mass divided by the volume. The mass is given in grams, and the volume can be expressed in cubic centimeters, millimeters, or liters. The units for density are therefore either grams/cubic centimeter (g/cm^3), grams/milliliter (g/mL), or grams/liter (g/L).

COMMON PITFALLS TO AVOID

Do not forget to check the logic of your answer, especially with unit conversion problems. If you measure your height in centimeters and after converting to feet, get 18 ft as your answer, you did something wrong. Look to see if your answer makes sense; should it be smaller or larger than the original number? If one pound is less than one kilogram then if you convert 145 lbs. to kg the number should be less than 145. One method of double-checking your answer is to take the answer you calculated and independently do the problem in reverse to see if you get the original number.

Many students learn the prefixes for the metric units but then get confused when actually using them in a conversion. For example, to convert from grams (g) to milligrams (mg), they know milli means 10^{-3} (or 0.001) but they forget whether $1\ g = 10^{-3}\ mg$ or $1\ mg = 10^{-3}\ g$. Again, think it through logically. If you know a mg is smaller than a gram, then "1" of the smaller unit must be equal to a really small portion of the larger unit (or it takes many small units to equal one larger unit). Therefore $1\ mg = 10^{-3}\ g$, i.e., $1\ mg = 0.001\ g$, or $1000\ mg = 1\ g$.

Keep track of units in a calculation and whether they are numerators or denominators. You should be able to see clearly which units get cancelled and which are left at the end of the calculation. For example, please remember that $g/(g/cm^3)$ will give you units of cm^3.

Be careful when rounding off numbers, e.g. rounding 1846 to two significant figures (it is not 18). Think of this number as though it were money. If you were owed $1846, would you want it rounded off to $18 or $1800? The trailing zeroes are necessary to show placement of the decimal point while still showing those digits are not significant. Do not do sequential rounding i.e., do not round 1846 to 1850 and then 1900. Simply look at the third digit to determine the value of the 2nd digit when rounding to 2 significant figures.

When carrying out a calculation stepwise be careful if you have a habit of rounding off at each step. (Check with your instructor to find out their policy.) In general it is better and more precise to carry a couple extra significant figures to the end of the calculation and save the full rounding off to the end of the calculation.

RECAP SECTION

This chapter is heavy on the math; it lays the foundation for the calculations you will need to master in order to be successful in the remaining weeks of the course. You should now be comfortable with converting from decimal format to scientific notation and vice versa. Understanding the precision to which to read measurements is important and the role and number of significant figures in both raw data and calculated results. Familiarity with the metric system, and the units for mass, length, and volume is necessary for your success in any scientific course or career. Being able to convert between English and metric units, and between the three temperature scales is a must. This chapter also presents strategies for solving calculation problems, in particular, the dimensional analysis approach. The concepts and applications of density will be utilized time and again.

SELF-EVALUATION SECTION

1. In order to handle large numbers efficiently, it is worthwhile to use scientific notation, which means writing a number as a power of 10. The numbers are always written between 1 and 10 with the associated power of 10.

    (1) Express the following numbers as powers of 10:

    0.0305 _____

    29 _____

    0.000721 _____

    680,000,000 _____

    36,800 _____

(2) Express the following in decimal form:

4.77×10^4 _____

8.41×10^{-2} _____

5.8×10^1 _____

9.1×10^0 _____

1.415×10^2 _____

2. Express the following numbers in exponential form:

(1) 275,000 _____ (4) 0.7920 _____

(2) 0.000361 _____ (5) 683000 _____

(3) 20,000,000 _____ (6) 40. _____

3. Express the following numbers in decimal form:

(1) 4.8×10^3 _____ (4) 1074×10^{-2} _____

(2) 0.67×10^2 _____ (5) 0.0034×10^3 _____

(3) 0.151×10^{-5} _____ (6) 11×10^{-5} _____

4. Round off the following numbers to four significant digits:

(1) 67.9840 _____ (5) 1.67049 _____

(2) 729.9998 _____ (6) 850,943 _____

(3) 0.0057432 _____ (7) 0.14951 _____

(4) 3,760,011 _____ (8) 0.03004 _____

5. (1) To what significant digit can you read each of the following volume measuring devices (for example, something could be read to the nearest 0.01 mL)?

(2) For each measuring device, read the volume of the liquid shown (all are in mL).

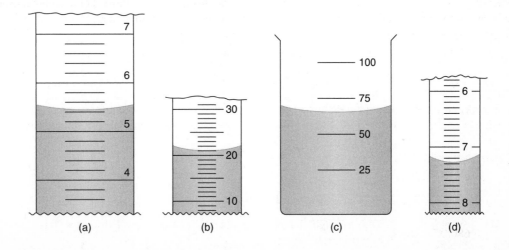

(a)   (b)   (c)   (d)

— 9 —

6. Three quantities routinely measured include (1) _____, (2) _____, and (3) _____. These basic measurements allow us to calculate many other quantities such as density which is (4) _____ /(5) _____.

The kilogram, abbreviated (6) _____, is used to measure (7) _____.

(8) _____ is measured in (9) _____, abbreviated m.

However, probably the most common type of measurement we make in a chemistry laboratory involves the metric volume (10) _____, which is abbreviated (11) _____.

To these three units of measure for mass, length, and volume, we can attach prefixes to change the unit's value by various powers of 10. For example, *kilo* means (12) _____ and *milli* means (13) _____.

The prefix for 0.01 (10^{-2}) is (14) _____, and the prefix for 0.000001 (10^{-6}) is (15) _____.

The abbreviations for the pre-fixes *kilo, centi, micro,* and *milli* are (16) _____, (17) _____, (18) _____, and (19) _____.

Using the four prefixes in general use in chemistry, we must be able to convert from one metric unit to any other corresponding metric unit. We will use a series of conversions to find out if a series of decimal-point changes presents any problems to you.

Convert These Units	To These Units
(20) 0.754 kg	_____ g
(21) 4.7 cm	_____ m
(22) 1.2 L	_____ mL
(23) 17.4 mg	_____ g
(24) 1005 μL	_____ mL
(25) 15 μg	_____ mg

Convert These Units	To These Units
(26) 0.25 g	_____ mg
(27) 0.34 km	_____ cm
(28) 1.25 mg	_____ g
(29) 0.0089 g	_____ mg
(30) 9 cm	_____ μm

7. This is a brief math review. Solve for x.

   (1) $x = (1.3 \times 10^2)(6.4 \times 10^{-1})$

   (2) $x = (5.76 \times 10^{-4})(2.4 \times 10^3)$

   (3) $(8.9 \times 10^1)\, x = (5.46 \times 10^3)$

   (4) $(0.77 \times 10^4)\, x = 125{,}000$

8. Converting units.

   (1) What is the difference in height in centimeters between an Olympic gymnast who is 1.47 m tall and an NBA player who is 6 ft. 7.0 in. tall?

Do your calculations here.

(2) Which soda is cheaper per oz (or mL), a 20. fl. oz grape soda for $0.90 or a 355 mL orange soda for $.75? What is the cost in cents per mL of the grape soda? One ounce equals 29.6 mL.

The words "cents per mL" are a way of expressing the fraction (cents/mL). That means that the number of cents is divided by the number of mL. The word "per" is a common method used in science to indicate a division operation or a fraction.

Do your calculations here.

(3) A Sacagawea dollar coin has an average mass of 8.10 grams and is 0.0857 ounces heavier than a quarter minted in the early 2000's. What is the average mass in grams of a quarter? 1 pound equals 453.6 g.

Do your calculations here.

9. We can convert the Kelvin temperature scale to the Celsius scale, and vice versa, with the formula $K = °C + 273.15$. The value 273.15 is a constant value.

Celsius degrees can then be converted to °F, or vice versa, using the formula

$$°F = (1.8 × °C) + 32 \qquad \text{or} \qquad °C = (°F − 32)/1.8$$

Using these formulas, make the following conversions.

(1) 273°F to _____°C

(2) 412 K to _____°C

(3) 25°C to _____°F

(4) 25°C to _____°K

(5) −28°F to _____°C

(6) Which was colder, a town in Siberia experiencing an unusually cold day of −62°C, or a research station in Antarctica, where the temperature was a chilly −75°F?

Do your calculations here.

(7) In Death Valley, CA, the highest temperature recorded was 57°C. By how many degrees Fahrenheit was that a change from the temperature in the research station in Antarctica?

Do your calculations here.

(8) Surface temperatures on Venus are upwards of 450°C. What is that temperature in °F?

Do your calculations here.

10. As you progress in your study of various substances, you will find a great diversity in the physical nature of matter. There are many ways to categorize matter in trying to relate the properties of one substance to another. One such way is through a relationship called *density*. The density of a substance is not only related to its mass but also to the volume it occupies. Thus, a cube of aluminum metal measuring 10 cm on a side has much less mass than a cube of iron metal of the same volume. We say the density of aluminum is less than that of iron.

The equation defining density is written as follows:

$$\text{Density} = \frac{\text{mass}}{\text{valume}}$$
$$d = \frac{\text{g}}{\text{cm}^3} = \frac{\text{g}}{\text{mL}} \text{ or } \frac{\text{g}}{\text{liter}}$$

The units for density vary depending on the units used for volume.

(1) Sample problem; Our block of aluminum, 10 cm on a side, was found to have a mass of 2700 g by using a balance. What is the density of aluminum?

Do your calculations here.

(2) Another problem: Logs of black ironwood, which do not float in water, are found to have a density of 1.077 g/cm^3. What is the volume of a piece of ironwood that has amass of 750 g?

Do your calculations here.

Challenge Problems

11. The standard of metric length is the meter which was defined (prior to 1983) as 1,650,763.73 wavelengths of a spectral line of the element krypton. To three significant figures determine this wavelength in nanometers if 1 nm = 10^{-9} m.

Do your calculations here.

12. For each of the following, identify (a) what is the numerical value and what is the unit, (b) how many significant figures, (c) which numbers are known and which are estimated. Round each number to two significant figures (d), and (e), express that number as a power of 10.

   (1)   1055 kg

   (2)   1,650,763 wavelengths

   (3)   1.077 g/cm^3

   (4)   0.00845 L

   (5)   776,000 g

ANSWERS TO QUESTIONS AND SOLUTIONS TO PROBLEMS

1. (1) 3.05×10^{-2}; 2.9×10^{1}; 7.21×10^{-4}; 6.8×10^{4}; 3.68×10^{4}
    (2)   47,700;   0.0841;   58;   9.1;   141.5

2. (1) 2.75×10^{5} (4) 7.920×10^{-1}
    (2) 3.61×10^{-4} (5) 6.83×10^{5}
    (3) 2×10^{7} (6) 4.0×10^{1}

3. (1)   4800   (4)   10.74
    (2)   67   (5)   3.4
    (3)   0.00000151   (6)   0.00011

4. (1)   67.98   (4) 3.760×10^{6} (7)   0.1495
    (2)   730.0   (5)   1.670   (8)   0.03004
    (3)   0.005743   (6)   850,900

5. (1)   (a)   nearest 0.05 (you are sure of the first decimal place, as the markings are every 0.2 mL; the spaces between the markings can be mentally divided into 4 equal portions, thus allowing you to estimate to the nearest 0.05)
         (b)   nearest 0.5
         (c)   nearest 1 (or if that is too hard, the nearest 2 or 5)
         (d)   nearest 0.05, or if really good at estimating, the nearest 0.02)

    (2)   (a)   5.45 mL
         (b)   21.0 (must include the .0 to show you read the volume to that precision)
         (c)   67 mL (66 or 68 is fine too)
         (d)   7.25 mL

6. (1), (2), (3) length, mass, volume (in any order) (4) mass, (5) volume, (6) kg, (7) mass, (8) length, (9) meters, (10) liter, (11) L (12) 1000 (13) 0.001 (14) centi- (15) micro- (16) k (17) c (18) μ (19) m

    (20)   754 g $(0.754 \, \text{kg}) \left(\dfrac{1000 \, \text{g}}{1 \, \text{kg}} \right) = 754 \, \text{g}$

    (21)   0.047 m $(4.7 \, \text{cm}) \left(\dfrac{1 \, \text{m}}{100 \, \text{cm}} \right) = 0.047 \, \text{m}$

(22)   1200 mL $(1.2\,\text{L})\left(\dfrac{1000\,\text{mL}}{1\,\text{L}}\right) = 1200\,\text{mL}$

(23)   0.0174 g $(17.4\,\text{mg})\left(\dfrac{1\,\text{g}}{1000\,\text{mg}}\right) = 0.0174\,\text{g}$

(24)   1.005 mL $(1005\,\mu\text{L})\left(\dfrac{1\,\text{mL}}{1000\,\mu\text{L}}\right) = 1.005\,\text{mL}$

(25)   0.015 mg $(15\,\mu\text{g})\left(\dfrac{1\,\text{mg}}{1000\,\mu\text{g}}\right) = 0.015\,\text{mg}$

(26)   250 $(0.25\,\text{g})\left(\dfrac{1000\,\text{mg}}{1\,\text{g}}\right) = 250\,\text{mg}$

(27) 3.4×10^4 cm $(0.34\,\text{km})\left(\dfrac{1000\,\text{m}}{1\,\text{km}}\right)\left(\dfrac{100\,\text{cm}}{1\,\text{m}}\right) = 34{,}000\,\text{cm} = 3.4 \times 10^4\,\text{cm}$

(28)   0.00125 g $(1.25\,\text{mg})\left(\dfrac{1\,\text{g}}{1000\,\text{mg}}\right) = 0.00125\,\text{g}$

(29)   8.9 mg $(0.0089\,\text{g})\left(\dfrac{1000\,\text{mg}}{1\,\text{g}}\right) = 8.9\,\text{mg}$

(30) 9×10^4 μm $(9\,\text{cm})\left(\dfrac{1\,\text{m}}{100\,\text{cm}}\right)\left(\dfrac{10^6\,\mu\text{m}}{\text{m}}\right) = 9 \times 10^4\,\mu\text{m}$

7.   (1) $x = 83$
     (2) $x = 1.4$
     (3) $x = 61$
     (4) $x = 16$

8.   (1)   The difference in height is 54 cm
           Given: 1.47 m tall and 6 ft 7.0 in. tall
           Knowns: $\dfrac{1\,\text{ft}}{12\,\text{in}}$ OR $\dfrac{12\,\text{in}}{1\,\text{ft}}$; $\dfrac{2.54\,\text{cm (exactly)}}{1\,\text{inch}}$ OR $\dfrac{1\,\text{inch}}{2.54\,\text{cm}}$; $\dfrac{100\,\text{cm}}{1\,\text{m}}$ OR $\dfrac{1\,\text{m}}{100\,\text{cm}}$

           Solution: convert both heights to centimeters and find the difference
                 basketball player: 6 ft 7.0 in = 6(12) + 7.0 in = 79.0 inches
           $79.0\,\text{in}\left(\dfrac{2.54\,\text{cm}}{1\,\text{in}}\right) = 201\,\text{cm}$

           gymnast: $(1.47\,\text{m})\left(\dfrac{100\,\text{cm}}{1\,\text{m}}\right) = 147\,\text{cm}$

           difference: 201 cm − 147 cm = 54 cm

     (2)   The cheaper soda is the grape soda.
           Given: 20. fl. oz grape soda for $0.90
                 grape soda costs (90 cents/20, fl. oz) = 4.5 cents per oz.
                 orange soda costs (75 cents/355 mL) = 0.21 cents per mL

           Knowns: $\dfrac{1\,\text{oz.}}{29.6\,\text{mL}}$ OR $\dfrac{29.6\,\text{mL}}{1\,\text{oz}}$

           Therefore: $\left(\dfrac{4.5\,\text{cents}}{\text{oz}}\right)\left(\dfrac{\text{oz}}{29.6\,\text{mL}}\right) = 0.15\,\text{cents per mL}$

     (3)   The average mass of a quarter is 5.67 g

           Given: dollar coin mass is 8.10 g

mass difference = 0.0857 ounces

Known: $\dfrac{453.6\,\text{g}}{1\,\text{lb}}$ OR $\dfrac{1\,\text{lb}}{453.6\,\text{g}}$; $\dfrac{1\,\text{lb}}{16\,\text{ounces}}$ OR $\dfrac{16\,\text{ounces}}{1\,\text{lb}}$

Therefore: $(0.0857\ \cancel{\text{ounces}})\left(\dfrac{1\ \cancel{\text{lb}}}{16\ \cancel{\text{ounces}}}\right)\left(\dfrac{453.6\,\text{g}}{1\ \cancel{\text{lb}}}\right) = 2.43\,\text{g}$

The difference in mass is 2.43 g (to 3 significant figures), so the average mass of the quarter is: 8.10 g − 2.43 g = 5.67 g

9.　(1)　134K　　(273°F − 32)/(1.8) = 134°C
　　(2)　139°C　(412 − 273) = 139°C
　　(3)　77°F　　(25°C)(1.8) − 32 = 77°F
　　(4)　298K　　(25°C + 273) = 298K
　　(5)　−33°C　(−28°F − 32)/(1.8) = −33°C

Reminder: significant figures in temperature conversions generally follow the same precision as the original measurement. So if the original temperature was measured to the nearest degree, the converted temperature is reported to the nearest degree.

　　(6)　the town in Siberia was colder
　　　　either convert −62°C to °F.
　　　　(−62°C)(1.8) + 32 = −79.6°F = −80.°F (sig. figs.); colder than Antarctica at −75°F
　　　　OR convert −75°F to °C
　　　　(−75°F − 32)/1.8 = −59°C; not as cold as the town in Siberia at −62°C that day.

　　(7)　210. degrees F difference
　　　　convert Death Valley's 57°C to °F, then find the difference:
　　　　(57°C)(1.8) + 32 = 134.6 °F − 135°F
　　　　difference of: 135°F − (−75°F) = 210. degrees

　　(8)　°F = (450°C × 1.8) + 32 = 842°F = 840°F

10.　(1)　2.7 g/cm^3

The mass is given as 2700 g, and we need to determine the volume of a cube 10 cm on a side. Volume is length × width × height. Multiplying 10 cm × 10 cm × 10 cm equals 1000 cm^3. Substituting into our equation, we have

$$d = \frac{\text{mass}}{\text{volume}} = \frac{2700\,\text{g}}{1000\,\text{cm}^3} = 2.7\,\frac{\text{g}}{\text{cm}^3}$$

The answer says that aluminum metal has a mass of 2.7 g per each cm^3. Therefore, 2 cm^3 would have a mass of 5.4 g, and so forth.

(2)　696 cm^3

We are given the density of the wood and the mass and asked to solve for volume. We write down the equation, then isolate the unknown quantity and take a close look at how the units of a problem can help us in solving equations.

$$d = \frac{\text{mass}}{\text{volume}}$$

Multiply each side by volume.

$$\text{volume} \times d = \frac{\text{mass} \times \cancel{\text{volume}}}{\cancel{\text{volume}}}$$

Now divide each side by density.

$$\frac{\text{volume} \times d}{d} = \frac{\text{mass}}{d}$$

Substitute our values into the modified equation.

$$\text{volume} = 750 \ \cancel{g} \times \frac{1 \ cm^3}{1.077 \ \cancel{g}}$$

$$= 696 \ cm^3 \ (3 \text{ significant figures})$$

Notice that "g" cancels out then leaves our answer with the units for volume, which are the desired units.

11. From the information given, if $1 \ nm = 10^{-9} \ m$, then $1 \ m = 10^9 \ nm$. Therefore, we can say that

$$1 \ m = 10^9 \ nm = 1{,}650{,}763.73 \text{ wavelengths}$$

To find the length of one wavelength we can set up the following two relationships:

$$10^9 \ nm = 1{,}659{,}763.73 \text{ wavelengths}$$

$$\frac{10^9 \ nm}{1{,}650{,}763.73} = 1 \text{ wavelength (dividing each side by } 1{,}650{,}763.73)$$

$$606 \ nm = 1 \text{ wavelength (to three significant figures)}$$

12. (1) 1055 kg: (a) 1055 is the numerical value, kg is the unit (b) 4 significant figures (c) 105 are known, last 5 estimated (d) 1100 kg to 2 significant figures (e) 1.1×10^3 kg

   (2) 1,650,763 wavelengths: (a) 1,650,763 is the numerical value, wavelengths are the units (b) 7 significant figures (c) 165076 are known, last 3 is estimated (d) 1,700,000 wavelengths to 2 significant figures (e) 1.7×10^6 wavelengths

   (3) 1.077 g/cm^3: (a) 1.077 is the numerical value, g/cm^3 are the units (b) 4 significant figures (c) 1.07 known, last 7 estimated (d) 1.100 g/cm^3 to 2 significant figures (e) 1.1×10^0 g/cm^3

   (4) 0.00845 L (a) 0.00845 is the numerical value, L is the unit (b) 3 significant figures (c) 84 known, 5 estimated (d) 0.0085 L to 2 significant figures (e) 8.5×10^{-3} L

   (5) 776,000 g (a) 776,000 is the numerical value, g is unit (b) 3 significant figures (c) 77 known, 6 estimated (d) 780,000 g to two significant figures (e) 7.8×10^5 g

CHAPTER THREE

Elements and Compounds

SELECTED CONCEPTS REVISITED

The case of the symbols used is exceedingly important especially if you turn in hand-written work. For example, CA, Cᴀ, and ca are unacceptable as a symbol for calcium. The correct symbol is Ca. Using the appropriate sizes and cases (upper versus lower) of letters is crucial.

When writing chemical formula for a compound, only the first letter of each elemental symbol is capitalized. Subscripts are written to the right of the elemental symbol. No spaces are left between elemental symbols making up a single compound.

Many instructors provide you with a periodic table for use during quizzes, tests etc. Although the periodic table shows symbols, most tables do not have the names of the elements. You will most likely have to learn the names of the most common ones. A couple symbols that students seem to get backwards often are P and K. P is phosphorus, K is potassium. The more you use these symbols, the faster you will learn the names. See #6 in the self-evaluation section for commonly used or confusing elements.

Being able to identify metals vs. metalloids vs. nonmetals will be invaluable as you learn more about the reactivity of elements and compounds. Remember that the first few chapters of the text form the foundation of your chemistry knowledge.

Elements found in nature are not necessarily found in their elemental form. For example, although gold is often found in a relatively pure state, aluminum is generally found as a part of a compound (or compounds) that must be chemically converted to pure elemental aluminum.

The Law of Definite Composition states that a compound has two or more elements in a definite proportion by mass. The Law of Multiple Proportions states that atoms of two or more elements may combine in different ratios to produce more than one compound.

COMMON PITFALLS TO AVOID

It is important that subscripts are written as such and that parentheses are in the appropriate position relative to the subscripts. $CuCN$ is not the same as $CuCN_2$ which is not the same as $Cu(CN)_2$.

Double-check all chemical formula in both typewritten and handwritten work. When writing or typing Co, did you mean Co for Cobalt, or CO for carbon monoxide? Is PB referring to lead (Pb) or phosphorus and boron? Is Mɴ supposed to be Mn for manganese or MN for a metal-nitrogen compound? Be aware that word-processing programs often automatically change the second letter of a word that has been capitalized back to lower case letters. Spell-check programs do not usually recognize chemical symbols.

RECAP SECTION

This chapter has introduced you to the elements and the Periodic Table. You should now be familiar with chemical symbols and it would be beneficial for you to be able to match some of those symbols with the element names. The Periodic Table imparts a great deal of information, much of it relating to the arrangement of the elements in the table. The separation into metals, nonmetals and metalloids, and the resulting prediction as to the type of compound formed, ionic or molecular, is material you will revisit constantly throughout this course. Knowing how to write chemical formula lays the groundwork for being able to represent chemical reactions later in the semester. In addition, chemical formulas describe the chemical composition of compounds, and follow the natural laws of definite composition and multiple proportions.

1. Fill in the missing word(s) from each sentence.

    The building blocks of all substances are called (1) _____; at least 88 occur naturally on earth.
    (2) _____ are composed of two or more elements chemically combined.
    When an element is subdivided into smaller and smaller pieces, the final indivisible particle that is left is called an
    (3) _____.
    When a compound, such as sugar, is subdivided into smaller and smaller pieces, the final indivisible particle that
    retains the identity of sugar, is called a (4) _____.
    Some compounds are formed from electrically charged species called (5) _____.
    Positively charged ions are called (6) _____ and negatively charged ions are called (7) _____.

2. Fill in the blank spaces.
    The three subgroups of elements are the metals, (1) _____, (2) _____.
    With the exception of (3) _____, all metals are in the solid state at room temperature.
    Common characteristics of metals include that they are good conductors of (4) _____ and
    (5) _____ and usually appear (6) _____.
    Metals are also (7) _____, which means they can be drawn into wires, and are (8) _____, which
    means they can be flattened into sheets.
    Most metals have (9) _____ melting points and (10) _____ densities.
    Silicon and arsenic are examples of (11) _____.
    Metals tend to react mainly with (12) _____ elements while nonmetals will react with any of the three
    subgroups.

3. Write the chemical symbol for the five most abundant elements in the earth's crust, seawater, and atmosphere. Also
    write the six most abundant elements in the human body. Which elements are common to both lists?

    Earth's Chemicals   **Human Body**

    a. _____   a. _____

    b. _____   b. _____

    c. _____   c. _____

    d. _____   d. _____

    e. _____   e. _____

                                             f. _____

    Elements common to both:

4. True or False

    (1) Atoms can combine only in simple whole number ratios.

    (2) Ionic compounds can be broken down into molecules.

    (3) There are over 100 known elements.

    (4) A chemical compound retains the identities of its component elements.

5. All of the following <u>elemental</u> symbols are incorrect. Indicate why they are incorrect and write the correct symbol.

    cad _____   CO _____

    NI _____   NA _____

mG _____ si _____

S_N _____ CU _____

6. Even if memorizing the periodic table is not required, knowing the symbols for some of the more common elements is useful. Please write the chemical symbol or name as appropriate for each element below.

Sn _____   S _____   _____ Silver

Na _____   N _____   Ni _____

Mn _____   Mg _____   _____ Mercury

Pb _____   _____ Potassium   _____ Phosphorus

Ca _____   _____ Cadmium   _____ Carbon

Ba _____   B _____   Br _____

_____ Copper   _____ Cobalt   _____ Gold

7. Each of the following pairs of elements belongs to a special sub-group on the periodic chart. What are the names of those special sub-groups?

Ar, Kr _____   Mg, Ba _____

Si, As _____   Cr, Rh _____

Li, K _____   Br, Cl _____

8. How many atoms of nitrogen are contained in each of the following formulas?

(1) $Na_3Co(NO_2)_6$ _____

(2) $NH_4S_2O_8$ _____

(3) $NH_4C_2H_3O_2$ _____

(4) NH_4CNS _____

(5) $Na_2Fe(CN)_5NO \cdot 2\,H_2O$ _____

(6) $K_3Fe(CN)_6$ _____

(7) $Fe_2(SO_4)_3(NH_4)_2SO_4 \cdot 24\,H_2O$ _____

9. (1) Name the appropriate columns or sections of the periodic table labeled below.

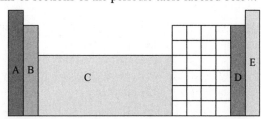

(2) Draw in the line separating the metals from the nonmetals.

(3) Which common element, although sitting on the dividing line, is considered fully a metal, and not a metalloid?

10. Look at the following list of chemicals. From what you've learned in Chapter 3, try to identify each formula as to whether it represents a compound or not, and if the formula is that of a compound whether the compound is molecular or ionic in nature as described in Chapter 3.

Formula	Compound? Yes or No	If yes then Molecular	or	Ionic
(1) Br_2	_____		_____	_____
(2) CO_2	_____		_____	_____
(3) S_8	_____		_____	_____
(4) NH_3	_____		_____	_____
(5) H_2O	_____		_____	_____
(6) CCl_4	_____		_____	_____
(7) Pb	_____		_____	_____
(8) KCl	_____		_____	_____
(9) Ar	_____		_____	_____
(10) NaI	_____		_____	_____

11. (1) What is unique about bromine? _____
    (2) What is unique about mercury? _____
    (3) Which diatomic element is a solid at room temperature? _____
    (4) Which picture is most likely to represent a diatomic element? a noble gas? _____

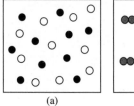
(a)
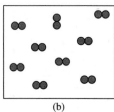
(b)
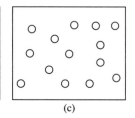
(c)

ANSWERS TO QUESTIONS AND SOLUTIONS TO PROBLEMS

1. (1) elements   (2) Compounds   (3) atom
   (4) molecule   (5) ions   (6) cations
   (7) anions

2. (1), (2) metalloids, nonmetals   (3) mercury
   (4), (5) heat, electricity   (6) lustrous or shiny
   (7) ductile   (8) malleable   (9) very high
   (10) high   (11) metalloids   (12) nonmetal

3. Earth: O, Si, Al, Fe, Ca
   Body: O, C, H, N, Ca, P
   Oxygen and calcium are common to both.

4. (1) True   (2) False   (3) True   (4) False

5. cad – elemental symbols are two only letters, with the first being capitalized – Cd
   NI – second letter should be lower case – Ni
   mG – first letter should be capitalized, second should be lower case – Mg
   Sɴ – second letter should be lower case, not a smaller upper case – Sn
   CO – second letter should be lower case – Co; CO implies carbon monoxide, not cobalt
   Nᴀ – second letter should be lower case, not a smaller upper case – Na
   si – first letter should be capitalized – Si
   CU – second letter should be lower case – Cu

6.
tin	sulfur	Ag
sodium	nitrogen	nickel
manganese	magnesium	Hg
lead	K	P
calcium	Cd	C
barium	boron	bromine
Cu	Co	Au

7.
noble gases	alkaline earth metals
metalloids	transition metals
alkali metals	halogens

8. (1) 6   (2) 1   (3) 1   (4) 2   (5) 6   (6) 6   (7) 2

9. (1)   A-alkali metals   B-alkaline earth metals   C-transition metals   D-halogens   E-noble/inert/rare gases

   (2)

   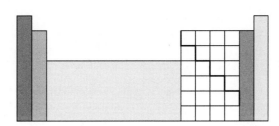

   (3)   Al

10. (1) No   (2) Yes, molecular   (3) No   (4) Yes, molecular   (5) Yes, molecular
    (6) Yes, molecular   (7) No   (8) Yes, ionic   (9) No   (10) Yes, molecular

11. (1)   Bromine is the only non-metal element that is a liquid at room temperature
    (2)   Mercury is the only metal element that is a liquid at room temperature
    (3)   Iodine
    (4)   (b) - diatomic, (c) - noble gas

CHAPTER FOUR

Properties of Matter

SELECTED CONCEPTS REVISITED

Physical properties can be determined without altering the substance, while chemical properties involve the interaction of substances resulting in different material.

A chemical change results in a new substance. Often you will see physical changes accompanying a chemical change because the new substance will likely have different physical characteristics than the original compound. If only a physical change takes place, no new substance is formed.

Although not strictly correct, atoms and molecules are often represented by balls. Different colors and/or sizes represent different atoms or elements. If the balls overlap, it usually means the atoms are chemically combined representing a multi-atomic element or compound.

A few reminders about chemical equations (they describe chemical reactions) . . .

$$\text{reactants} \rightarrow \text{products}$$
$$(\text{starting substances} \rightarrow \text{new substance(s) formed})$$

Potential energy is the energy which matter has as a result of chemical bonds or because of its position in relationship to another place; kinetic energy is strictly the energy of motion.

When we talk about the amount of heat energy involved in a chemical process, we use the SI unit of measurement called the *joule*. We define the joule by an experimental procedure – a common practice in science. If we measure out 1 gram of water and place it at 14.5° Celsius, then 4.184 joules of heat energy from a match, gas burner, or other source will raise the temperature of the 1 gram of water 1° Celsius to 15.5° Celsius. In words, the number of joules gained or lost is equal to the mass of water times the temperature change times a constant. The constant is the *specific heat* for the material in question. The higher the specific heat, the more heat is needed to raise the temperature by one degree. Therefore, a substance with higher specific heat will need to absorb more energy to raise its temperature than a substance with a lower specific heat. For water, the specific heat is 4.184 J/g°C.

Remembering that the number of joules (abbreviated J) gained or lost during heat energy changes is equal to the mass of the substance times the specific heat of the substance times the temperature difference, we can write the following equation.

$$\text{joules} = (\text{grams of substance})(\text{specific heat of substance})(\Delta t)$$

The symbol Δt means "temperature difference". The next to the last term in the equation is the specific heat for the substance in question. For water, the specific heat is 4.184 J/g°C. The equation for water with units attached would be

$$\text{joules} = \text{g} \times 4.184 \, \text{J/g°C} \times °\text{C}$$

You have learned that heat is not the same as temperature. They are however inextricably linked. Heat flows between two bodies until the temperatures are equal, Therefore bodies at higher temperatures lose heat until they reach the same temperature as a body at a cooler temperature. This is not to say that a body at high temperature necessarily has more heat energy than a body at a cooler temperature since the amount of heat energy within the body is related to its mass.

COMMON PITFALLS TO AVOID

Do not forget to be consistent with your units! One of the most common mistakes students make is to forget to check that their units are consistent. You cannot add a number in Joules to a number in kilojoules without first converting one of the numbers to the same unit as the other.

It is best not to start to solve a problem without first reading through the entire problem. Here are some hints to help you solve problems. Remember when you approach problem solving you should always read the entire problem through once to get a quick feel for the topic and what you are being asked to solve. You should then reread the problem picking out the important information and organizing it in a way you can easily understand. For example, if a problem involves numbers such as "the mass of an object occupying 12 ml is 30.5 g", write the variable symbol followed by the given number for easy reference (e.g. m = 30.5 g, V = 12 ml). Try to determine what you are being asked for and where you can get the information you need, that is, is it given to you in the question, is it something you are supposed to have memorized, is it found on the periodic chart, etc?

Heat ≠ temperature; heat content depends on the mass of the object. Temperature is independent of mass.

RECAP SECTION

By the end of Chapter 4, you should be able to distinguish between chemical and physical properties of matter, and identify which type of property of a particular substance was being investigated. Often chemical and/or physical changes take place when investigating the properties of a substance. The chemical equation, used to show changes and whose basic format is introduced here, will be in recurring use throughout the course. Consistent application of problem solving techniques is covered and immediately put in practice for calculations involving the heat losses or gains for given systems. Energy, both kinetic and potential, its commonly used units of joules and calories, and the Law of Conservation of Energy are included in this chapter. Hydrocarbons as a major world source of energy are introduced here.

SELF-EVALUATION SECTION

1.   Determine whether the following statements or blanks relate to *chemical, light, thermal, mechanical,* or *electrical* energy.

When wood is burned the two forms of energy that are released are ___(1)___ and ___(2)___

(1) _____

(2) _____

The chemical energy of a car battery is converted into ___(3)___ energy, which turns the starter motor.

(3) _____

Fuels such as gasoline contain ___(4)___ energy which, through the process of combustion, is converted to the ___(5)___ energy of moving pistons, valves, gears, and wheels.

(4) _____

(5) _____

Hydroelectric dams use the energy of falling water to turn turbines that convert ___(6)___ energy into ___(7)___ energy.

(6) _____

(7) _____

Radiant energy ___(8)___ from the sun is used in the process of photosynthesis.

(8) _____

Photosynthesis involves a conversion of radiant energy into the ___(9)___ energy of sugars, starch, and other compounds.

(9) _____

Organisms utilize the ___(10)___ energy of food-stuffs for movement and activity and, in the case of warm-blooded animals, for maintaining body temperature ___(11)___.

(10) _____

(11) _____

2. Statements about the properties and changes of various substances are given below. Place the appropriate letter of identification in each space provided.

    a.   physical property   b.   chemical property

    c.   physical change   d.   chemical change

Description	**Identification**
(1) Potassium is solid at room temperature.	_____
(2) Potassium is among the most reactive elements in nature.	_____
(3) The melting point of potassium is 63.6 °C.	_____
(4) Potassium reacts vigorously with water.	_____
(5) Potassium added to water produces hydrogen gas plus a water solution of potassium hydroxide.	_____
(6) At 760 °C, potassium begins to boil and to change from a liquid to a gas.	_____
(7) At high temperatures, potassium reacts with hydrogen gas to form a metallic hydride.	_____
(8) Alkali metals, such as potassium, form salts with many other elements.	_____
(9) Potassium metal is shiny and fairly soft.	_____
(10) Potassium is useful in the manufacture of fertilizer compounds.	_____

3. The specific heats of iron, aluminum, and copper are 0.473, 0.900, and 0.385 J/g°C, respectively. Three chairs of similar mass, each made of one of the metals above, were placed in the sun. Based on their specific heats, which chair would you expect to warm up fastest? Explain briefly.

4. (1) How many grams of water can be heated from 5 °C to 61°C with 1.0×10^3 joules of thermal energy?

    Do your calculations here.

    (2) How many kilojoules of thermal energy would be required to heat a flask containing 180 mL and (180 g) of water from room temperature (25 °C) to 82 °C?

    Do your calculations here.

    (3) Let's work problems (1) and (2) again but this time we will use the calorie unit of heat measurement. Recall that the specific heat of water in calorie units is 1 cal/g°C or in words "one calorie of heat energy will raise one gram of water one degree Celsius". Therefore, how many kilocalories of thermal energy are required to heat 180 g of water from 25 °C to 82 °C?

    Do your calculations here.

    (4) In a similar manner, how many grams of water can be heated from 5 °C to 61 °C with 1.0×10^3 cal of thermal energy?

    Do your calculations here.

5. Under the right conditions ammonia can be made from nitrogen gas and hydrogen gas.

$$\text{nitrogen gas} + \text{hydrogen gas} \rightarrow \text{ammonia gas}$$

(1) What is (are) the reactant(s)? What is (are) the product(s)?

(2) If at the start of the reaction, 20.0 g each of hydrogen gas and nitrogen gas are used, and 15.7 g of hydrogen gas is left when the reaction is complete, what is the percent composition of N and H in ammonia? Assume the mass of nitrogen used plus the mass of hydrogen used equals the mass of ammonia produced.

Do your calculations here.

Challenge Problems

6. Three substances A, B, and C are undergoing investigation in the laboratory. The specific heat capacity of A is known to be 0.517 J/g°C.

(1) 10.0 g each of A and B are heated under identical conditions. After 5 minutes of heating, A is 3 °C warmer than B. Which substance has a higher specific heat, A or B? Briefly explain.

(2) The specific heat of C is found to be nearly twice that of A. Will 5.0 g of C or 5.0 g of A lose more energy when each sample drops 8 °C? Briefly explain.

7. A 50-g piece of copper at 212 °F is dropped into 100 g of ethyl alcohol at 72 °F. The specific heat of copper is 0.0921 $\frac{cal}{g°C}$, that of ethyl alcohol is 0.511 $\frac{cal}{g°C}$. Calculate the final temperature of the mixture.

Do your calculations here.

ANSWERS TO QUESTIONS AND SOLUTIONS TO PROBLEMS

1. (1) thermal   (2) light   (3) electrical   (4) chemical
   (5) mechanical   (6) mechanical   (7) electrical   (8) light
   (9) chemical   (10) mechanical   (11) thermal

2. (1) a   (2) b   (3) a   (4) b
   (5) d   (6) c   (7) d   (8) b
   (9) a   (10) b

3. The copper chair.
   Assuming all three absorbed the same amount of thermal energy, the biggest change in chair temperature should come from the chair with the lowest specific heat.

4. (1) 4.3 g. This problem is worked similarly to that in the text. The rest of this question uses a similar approach, although the solutions are not as detailed.

Knowns: initial temp $= 5\,°C$; final temp $= 61\,°C$
            $\Delta t = 61 - 5 = 56°C$
            specific heat of water $= 4.184\ J/g°C$
            heat in J $= 1.0 \times 10^3\ J$

solving for: mass
equation: heat in J $=$ mass $\times$ specific heat $\times \Delta t$

rearrange: $$\text{mass} = \frac{\text{heat in J}}{(\text{specific heat})(\Delta t)}$$

solve: $$\text{mass} = \frac{1.0 \times 10^3\ J}{(4.184\ J/g°C)(56\,°C)} = 4.267\ g = 4.3\ g\ (\text{to 2 significant figures})$$

(2) 43 kJ. We know the mass of water (180 g) and can determine the temperature difference by subtraction. The unknown quantity is the number of kilojoules of heat energy required. Specific heat has units involving joules so the heat energy is first found in joules and then divided by 1000 to give the answer in kilojoules (and two significant figures).

$$\text{joules} = m \times \text{specific heat} \times \Delta T$$
$$= (180\ g)(4.184\ J/g°C)(82 - 25)°C$$
$$= 42928\ J$$
$$= 43\ kJ$$

(3) 10. kcal. We know the mass of water (180 g) and know the temperature difference by subtraction. The unknown quantity again is the number of calories of heat energy required. Make the necessary substitutions in the equation.

$$\text{cal} = (180\ g)(1\ cal/g°C)(82\,°C - 25\,°C)$$
$$= (180\ g)(1\ cal/g°C)(57\,°C)$$
$$= 10260\ cal$$
$$\text{kcal} = 10.\ kcal\ (\text{2 significant figures})$$

(4) 18 g of water
Again, what mass of water can 1.0×10^3 cal raise from 5 °C to 61 °C?
Solving our equation for grams we will have

$$\text{cal} = m \times \text{specific heat} \times \Delta t$$
$$m = \frac{\text{cal}}{(\text{specific heat})(\Delta t)}$$
$$= \frac{1.0 \times 10^3\ cal}{(1\ cal/g°C)(61\,°C - 5\,°C)}$$
$$= \frac{1.0 \times 10^3\ cal}{(1\ cal/g°C)(56\,°C)}$$
$$= 18\ g\ \text{of water}$$

5. (1) reactants are nitrogen gas and oxygen gas
   (2) 82.3% N and 17.7% O
       solving for: % N and %H in ammonia
       knowns: 20.0 g nitrogen gas used
                       $20.0 - 15.7\ g = 4.3\ g$ of hydrogen gas used

$$\text{total mass of ammonia formed} = 20.0 + 4.3 = 24.3 \text{ g}$$

solution:

$$\%N = \frac{\text{mass nitrogen gas}}{\text{mass of ammonia}} = \frac{20.0 \text{ g}}{24.3 \text{ g}} \times 100 = 82.3\%$$

$$\%H = \frac{\text{mass hydrogen gas}}{\text{mass of ammonia}} = \frac{4.3 \text{ g}}{24.3 \text{ g}} \times 100 = 17.7\%$$

6. In examining the problem we should notice first that the temperatures are given in °F while specific heats contain the temperature unit of °C. In general, as we work with chemical problems we will find that conversions are necessary when we see temperatures given as °F. Therefore, the copper metal's temperature is

$$°C = \frac{(°F - 32)}{1.8} = \frac{(212 - 32)}{1.8} = 100 °C$$

The temperature of the ethyl alcohol is

$$°C = \frac{(°F - 32)}{1.8} = \frac{(72 - 32)}{1.8} = 22 °C$$

Next we can say that the copper metal piece has been heated up, dropped into a cooler material, the alcohol, and caused the final mixture to increase in temperature. The calories of heat gained by the copper as it was heated are lost to the alcohol so that the final temperature of the copper and alcohol are equal.

Another way to state this is that the heat lost by the copper is equal to the heat gained by the alcohol.

Mathematically, the equality would be:

$$\text{heat lost}_{Cu} = \text{heat gained}_{alcohol}$$

$$\text{heat lost}_{Cu} = m_{Cu}c_{Cu}\Delta T_{Cu} = (50 \text{ g})\left(0.092 \frac{\text{cal}}{\text{g°C}}\right)(100°C - T_f)$$

$$\text{Heat gained}_{alcohol} = m_{alcohol}c_{alcohol}\Delta T_{alcohol} = (100 \text{ g})\left(0.511 \frac{\text{cal}}{\text{g°C}}\right)(T_f - 22°C)$$

(c_{Cu} and $c_{alcohol}$ represent the specific heats of copper and alcohol respectively.)

Notice that T_f is second in ΔT_{Cu} and first in $\Delta T_{alcohol}$. This is a result of the Cu being cooled (to a lower temperature) while the alcohol is being warmed (to a higher temperature).

Since heat lost$_{Cu}$ = heat gained$_{alcohol}$

$$(50. \text{ g})\left(0.0921 \frac{\text{cal}}{\text{g°C}}\right)(100°C - T_f) = (100. \text{ g})\left(0.511 \frac{\text{cal}}{\text{g°C}}\right)(T_f - 22°C)$$

Upon expansion,

$$460.5 \text{ cal} - \left(4.605 \frac{\text{cal}}{°C}\right)T_f = \left(51.1 \frac{\text{cal}}{°C}\right)T_f - 1124.2 \text{ cal}$$

$$1584.7 \text{ cal} = \left(55.705 \frac{\text{cal}}{°C}\right)T_f$$

$$28 °C = T_f$$

7. (1) B has a higher specific heat than A. Substance B required more energy to raise its temperature by a degree so for the same amount of energy input, its overall temperature rise was smaller than that of substance A.

(2) Substance C will lose more energy than substance A. If C has a higher specific heat it means more energy is involved in changing the temperature of a sample by one degree. Therefore for similar temperature drops, C must have released nearly twice the amount of energy than A.

CHAPTER FIVE

Early Atomic Theory and Structure

SELECTED CONCEPTS REVISITED

The Law of Definite Composition states that a compound has two or more elements in a definite proportion by mass. The Law of Multiple Proportions states that atoms of two or more elements may combine in different ratios to produce more than one compound.

Two important features of Dalton's theory were later proved wrong by the existence of isotopes and by the discovery of subatomic particles. Atoms are organized as a small nucleus containing the protons and the neutrons. A cloud of electrons surrounds the positively charged nucleus. Electrons are the only subatomic particles involved in most reactions with the exception of nuclear reactions (covered in a later chapter). When electrons are lost or gained from an atom, the atom becomes charged and is referred to as an ion (a charged species). Because the protons and the neutrons have much greater mass than the electrons, we rarely distinguish between the mass of an ion versus the mass of the corresponding atom.

$$\begin{aligned} \text{mass } \#\ (p+n) &\rightarrow \text{A} \\ \text{atomic } \#\ (p) &\rightarrow \text{Z} \end{aligned} \quad \text{X}$$

The atomic mass unit (amu) is introduced in this chapter. The weight of a single atom is very small so scientists developed a new unit (amu) to avoid having to use numbers on the order of 10^{-24} on a regular basis. The numbers for the masses on the periodic chart when followed by the unit amu refer to the weight of single atoms. You will see in a later chapter that when grams are used with the periodic chart numbers, many more than single atoms are being counted/weighed. Amu's are not weighed in the lab.

A law describes behavior; a theory explains behavior.

COMMON PITFALLS TO AVOID

When it is appropriate to write the charge on an ion? When an ion is written by itself as opposed to as part of a compound, the charge must be shown as a superscript to the right of the symbol. Na is not the same as Na^+. When the ion is part of a compound the charges are generally not written, for example, we usually write NaCl not Na^+Cl^- (you may see exceptions to this used when we want to emphasize the charges).

If the charge on an ion is greater than 1, write the number before the sign, e.g., Mg^{2+} not Mg^{+2}. You will later learn that the number after the sign can mean something else (oxidation number).

Isotopes must have the same atomic number. $^{14}_{6}C$ and $^{14}_{7}N$ are not isotopes.

Isotope $\neq$ Ion

Isotope - #p same, # n different - i.e., mass numbers are different, atomic numbers are same
Ion - only #e changes - i.e., nucleus is unaffected, only the number of electrons is different

RECAP SECTION

With Chapter 5 now completed, you know much more about the atom. Although Dalton's atomic model assumed atoms were indivisible, subatomic particles, the proton, the neutron, and the electron, were discovered. Rutherford's modified atomic model incorporated the arrangement of these particles within the atom. Coulomb's Law is introduced and the role of electrical charge and its resulting force in atomic and chemical properties will be revisited often in future chapters. Atomic numbers, atomic masses, and isotopes, are all dependent on the numbers of subatomic particles in each atom. Identifying cations and anions, and recognizing the subatomic particle affected, the electron, is crucial to furthering your understanding of chemistry.

SELF-EVALUATION SECTION

1. Match the scientist to the proposal, discovery, or theory he made.

    Chadwick   _____(1) discovered radioactivity

    Becquerel   _____(2) alpha particle experiment results in proposal of a heavy, dense atomic nucleus

    Thomson   _____(3) discovered the neutron, a particle with mass but no charge

    Rutherford   _____(4) experimentally showed the existence of the electron; also demonstrated protons are
                            particles

    As a result of these and other experiments and discoveries, the three sub-atomic particles of interest are (5) _____, (6) _____, and (7) _____.

    This also led to two subsequent proposals for a model of an atom as illustrated below.
    (8) Which model was proposed first and by whom?

2. What are three isotopes of hydrogen? Give their common names and symbols.

3. Indicate by a yes or no response which of the following statements are generally accepted today.

    (1) Atoms of the same element may vary in mass but are the same size.
    (2) Compounds are formed by the union or joining of two or more atoms of different element. _____
    (3) Atoms of different elements have different masses and sizes._____
    (4) Atoms of two elements may combine in different ratios to form more than one compound. _____
    (5) Elements are composed of small, indivisible units called electrons._____
    (6) In forming compounds, atoms can combine in fractional units._____
    (7) The fundamental building blocks of elements are small particles called atoms._____
    (8) Atoms of the same element are alike in mass and size._____
    (9) In forming compounds, atoms combine in small whole-number ratios such as one to one, one to two, and so forth._____
    (10) Atoms of two elements can combine in only one fixed ratio to form only one compound._____

4. (1) This type of ion has a smaller diameter than its corresponding atom._____
    (2) This atom is from the same element as another atom but has a different mass._____
    (3) This is a negatively charged subatomic particle. _____
    (4) This English scientist realized the aqueous solutions of certain substances could conduct electricity. _____
    (5) This scientist developed an instrument, a tube, named after him, which generates emissions called cathode rays._____

5. Fill in the blank spaces.

(1)_____ number ———→ ${}^{7}_{3}\text{Li}$

(2)_____ number ———→

(1)_____ number = number of (3) _____ + number of (4) _____

(2) _____ number = number of (5) _____

In a neutral atom, the number of (6) _____ = (7) _____

6. Description of properties of subatomic particles. Fill in the spaces of the table with the correct response from the list of possible responses below.

Particle	Symbol	Relative Mass (amu)	Charge	Location in Atom
Electron				
Proton				
Neutron				

Possible responses

Symbol	Relative Mass (amu)	Charge	Location in Atom
	+2	−1	nucleus
p	+1	−2	outside nucleus
	−1	+1	
e	$-\frac{1}{2}$	−2	
n	0	0	
	$\frac{1}{1837}$	$-\frac{1}{2}$	

7. Atomic masses, atomic numbers, and number of neutrons.

(1) Determine the number of neutrons in an atom of each of the following elements.

${}^{23}_{11}\text{Na}$ _____ ${}^{65}_{30}\text{Zn}$ _____

${}^{115}_{48}\text{Cd}$ _____ ${}^{40}_{18}\text{Ar}$ _____

${}^{37}_{17}\text{Cl}$ _____ ${}^{24}_{12}\text{Mg}$ _____

(2) Determine the approximate *atomic mass* for an atom of each of the following isotopes.

	Atomic Number	Number of Neutrons	Atomic Mass
Li	3	4	_____
Al	13	14	_____
Br	35	46	_____
Fe	26	30	_____
H	1	0	_____
O	8	10	_____

8. Which of the following statements are true about the neutral atoms $^{12}_{6}C$ and $^{14}_{6}C$?

(1) $^{12}_{6}C$ and $^{14}_{6}C$ are isotopes.
(2) $^{12}_{6}C$ has two less electrons than $^{14}_{6}C$.
(3) $^{14}_{6}C$ has two more protons than $^{12}_{6}C$.
(4) $^{12}_{6}C$ has two less neutrons than $^{14}_{6}C$.

9. Given the following atomic mass units, decide which isotope is more abundant for each element below.

	Isotope A	Isotope B	Isotope C	avg amu
(1)	62.93 amu	64.93 amu	———	63.55 amu
(2)	49.95 amu	50.94 amu	———	50.94 amu
(3)	31.97 amu	32.97 amu	33.97 amu	32.065 amu
(4)	6.015 amu	7.016 amu	———	6.941 amu

10. Use the following diagrams to answer the questions below.

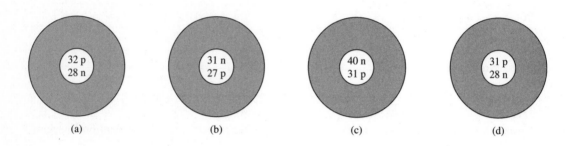

(a)   (b)   (c)   (d)

(1) Which pictures represent isotopes of each other?_____

(2) Given that these isotopes have masses of 68.925 amu (60.11%) and 70.925 amu (39.89%), calculate the average amu for this element.

(3) Name the element each picture represents._____ _____ _____ _____

Do your calculations here.

11. (a) The actual mass of a Carbon-12 atom is 1.9927×10^{-23} g. What is the relative atomic mass (i.e., the mass that would appear under the symbol of the element on the periodic table) of an element Jamcium whose mass is 4.0986×10^{-21} g?

(b) If there are only two isotopes of this element and they differ by only one neutron, which is more naturally abundant, the lighter or the heavier isotope? How do you know?

ANSWERS TO QUESTIONS AND SOLUTIONS TO PROBLEMS

1. (1) Becquerel   (2) Rutherford   (3) Chadwick   (4) Thomson
(5), (6), (7) proton, neutron, electron (in any order)
(8) Thomson proposed the model on the right first. After his gold foil experiment, Rutherford proposed the heavy, dense nucleus model of the atom.

2. hydrogen $^{1}_{1}H$ deuterium $^{2}_{1}H$ tritium $^{3}_{1}H$

3. (1) yes   (2) yes   (3) yes   (4) yes   (5) no
(6) no   (7) yes   (8) no   (9) yes   (10) no

4. (1) cation (electrons are lost so the remaining electrons are held more closely to the positive nucleus).
   (2) isotope   (3) electron
   (4) Faraday   (5) Crookes

5. (1) mass   (2) atomic   (3), (4) protons, neutrons
   (5) protons   (6), (7) protons, electrons

6.

Particle	Symbol	Relative Mass (amu)	Charge	Location in Atom
Electron	e	1/1837	−1	outside nucleus
Proton	p	1	+1	nucleus
Neutron	n	1	0	nucleus

7. (1)   Na = 12   Zn = 35
         Cd = 67   Ar = 22
         Cl = 20   Mg = 12
   (2)   Li = 7   Fe = 56
         Al = 27   H = 1
         Br = 81   O = 18

8. (1) T   (2) F   (3) F   (4) T

9. (1) Isotope A – the average amu is closer to the mass of the lighter isotope; with only two isotopes, the weighted average amu will be closer in mass to the more abundant isotope.
   (2) Isotope B   (3) Isotope A   (4) Isotope B

10. (1)   c, d

    (2)   69.72 amu
    Find the weighted average of the two isotopes. 60.11% of the mass comes from the lighter isotope and 39.89% comes from the heavier isotope. Find the contribution from each and the sum is the average amu.

$$(68.925 \text{ amu})(60.11/100) = 41.43 \text{ amu}$$

$$(70.925 \text{ amu})(39.89/100) = 28.29 \text{ amu}$$

$$\text{average mass} = 41.43 + 28.29 = 69.72 \text{ amu}$$

    (3)   (a) Ge   (b) Co   (c) Ga   (d) Ga

11. (a)   205.68

    (b)   The heavier isotope. Since the isotopes differ by only one mass number, the nuclides must weigh 205 and 206. If there were equal amounts of each isotope, the average mass would be 205.5. Since the average mass is greater than 205.5, the heavier isotope must be more abundant.

SELECTED CONCEPTS REVISITED

Naming compounds is an essential aspect of chemistry. Some names and their corresponding chemical formula you will simply have to memorize, such as the common names for certain compounds. Ask your instructor for the specific common names that will be required to know. Some of the more popular ones are laughing gas, N_2O, baking soda, $NaHCO_3$, and table salt, $NaCl$. You will also need to become familiar with many polyatomic ions. The oxygen-containing polyatomic ions often involve the root name of the non-oxygen component, for example, phosphate would contain P and O (PO_4^{3-}), and nitrite contains N and O (NO_2^-). The more you use the names and formula, the faster you will learn them. Get into the habit of trying to name all simple compounds you meet in the text or in class.

For ionic compounds, the name consists simply of the name of the cation (using a Roman numeral if necessary) followed by the name of the anion. (Think: cation anion)

For covalent compounds it is important to remember to use <u>prefixes</u> in the name to indicate the subscripts in the formula of the compound.

Compounds that are acids (the formula begins with H) can be named either as ionic compounds or specifically as acids.

The subscript in a formula indicates the number of atoms chemically bonded together to form that molecule or unit of a compound. Roman numerals are used only in the names of transition metal compounds, not in their formula, and indicate the cation's change. Roman numerals are NOT equal to the subscript of the metal.

Although most transition metals have more than one possible ion, it is important to understand in a particular compound only one ion exists at a time. That is, if you broke down the compound Mn_2O_5 both manganese ions are Mn^{5+} (*not* one Mn^{7+} and one Mn^{3+}).

Parentheses are necessary to show more than one polyatomic ion in a formula, e.g., $Mg(OH)_2$ means there are two OH^- ions.

COMMON PITFALLS TO AVOID

The subscript on a transition metal is not the same as the Roman numeral after the metal's name in the compound name. The subscript indicates the number of ions involved in a single unit of the compound while the Roman numeral indicates the *charge* on the metal ion. Roman numerals are only necessary when the transition metal has more than one possible ion.

chromium (VI) oxide
$\uparrow$ $\uparrow$ $\uparrow$ $\Rightarrow$ Cr^{6+}, O^{2-} $\Rightarrow$ CrO_3
Cr   6+ O^{2-}

diphosphorus pentoxide   nitric acid
$\uparrow$ $\uparrow$ $\uparrow$ $\uparrow$ $\Rightarrow$ P_2O_5 $\uparrow$ $\uparrow$ $\Rightarrow$ HNO_3
2   P   5   O NO_3^- H^+

Roman numerals are <u>not</u> written in the chemical formula of a compound.

Be careful when referring to ions, for example, do *not* refer to K^+ as a potassium *atom*. While this may seem to be nit picking, the behavior of a potassium atom is actually very different from the behavior of a potassium ion so it is very important that you refer to the different species correctly.

Trying to name compounds by simply naming every element is a common mistake made in early chemistry classes. Most compounds you meet at this level that are composed of more than two elements will most likely be ionic compounds and involve a polyatomic ion. You will not be able to name ionic compounds if you cannot recognize polyatomic ions. For example, K_2CO_3 is *not* potassium carbon oxide but *is* potassium carbonate.

Do not try to memorize the formula for ionic compounds. There are too many variations. You need to know the polyatomic ions and you should be able to look at the periodic table to determine the charge of the ions formed from certain elements. Once you know these you should be able to figure out the formula for the different ionic compounds. If you understand this it may make it easier for you to remember that you do not need prefixes (di, tri etc) in the names of ionic compounds to describe the subscripts because we can always determine the subscripts from the charges on the ions. In covalent compounds, if we are not told the subscripts in the name, we have no way of knowing the subscripts in the formula because the formula for the compound is not based on balancing charges.

Try not to memorize all the charges on monatomic ions. When you get to electronic structure of atoms you will be able to understand why certain atoms form specific ions but in the meantime use the periodic chart to help you. The text explains the trends. Remember you generally have access to a periodic chart.

RECAP SECTION

Chapter 6 is one of the self-contained study units in the text for you to refer to as needed. Although most compounds are now names systematically, a few common names remain in use, especially those chemicals that have high non-academic usage. Names of compounds depend on the type of compound, so understanding how ions are formed and interact becomes important. Familiarity and proficiency with writing names from formulas and formulas from names is essential. The types of compounds include ionic, binary covalent, and acidic. The textbook has a couple very nice flowcharts that help you walk through how to name compounds; Fig 6.4 for binary compounds, and 6.5 for acids. This is a chapter where practice and experience will be invaluable.

SELF-EVALUATION SECTION

This is a chapter more easily mastered with LOTS of practice!
1. Write formula for the following elements and ions. Recall that the charges must be written for the ions.

   (1) sulfur _____   (2) sulfide _____

   (3) lithium _____   (4) lithium ion _____

   (5) phosphorus _____   (6) phosphide _____

2. Write the chemical equation showing the loss or gain of electrons for the elements in number 1 above.

3. Write the correct formula for the ions first and then write the formula for the compound. (These are the steps you should always take when given the name of an ionic compound and asked to write the formula.)

   (1) aluminum ion _____ oxide _____ aluminum oxide _____

   (2) iron (III) ion _____ bromide _____ iron (III) bromide _____

(3)	ammonium ion	_____	sulfate	_____	ammonium sulfate	_____
(4)	magnesium ion	_____	phosphate	_____	magnesium phosphate	_____
(5)	copper (II) ion	_____	cyanide	_____	copper (II) cyanide	_____
(6)	potassium ion	_____	carbonate	_____	potassium carbonate	_____

4. Name the following ionic compounds. Note that the process has been broken down to give you extra practice.

Formula	cation symbol	cation name	Roman numeral?	anion symbol	anion name	COMPOUND NAME
KNO_3	_____	_____	_____	_____	_____	_____
$SnCl_4$	_____	_____	_____	_____	_____	_____
Na_2SO_4	_____	_____	_____	_____	_____	_____
$K_2Cr_2O_7$	_____	_____	_____	_____	_____	_____
$FeCl_3$	_____	_____	_____	_____	_____	_____
PbO_2	_____	_____	_____	_____	_____	_____
NH_4NO_3	_____	_____	_____	_____	_____	_____

5. Now, let us take a similar step-wise approach to naming covalent compounds. Name the following covalent compounds.

Formula	prefix	name of 1st nonmetal	prefix	name of 2nd nonmetal	NAME
SO_3	_____	_____	_____	_____	_____
Si_2Cl_6	_____	_____	_____	_____	_____
IBr	_____	_____	_____	_____	_____
NO_2	_____	_____	_____	_____	_____

6. On the following list of formula,
   (1) Circle the acids   (2) Underline the non-metallic binary compounds

HCl	CO	SnO_2	Al_2O_3	PBr_3	$PbCl_2$
P_2O_5	H_2CO_3	$NaCl$	$MgCl_2$	H_2SO_4	N_2O_3

7. Write the names for the formulas and the formulas for the written names listed below. Refer to the tables in Chapter 6 of the text book.

(1) $AlPO_4$ _____

(2) $(NH_4)_2CO_3$ _____

(3) NaI _____

(4) BaO _____

(5) NaHCO$_3$ _____

(6) Hydrochloric acid _____

(7) Lead (II) oxide _____

(8) Potassium permanganate _____

(9) Magnesium chloride _____

(10) Silicon dioxide _____

(11) Sulfuric acid _____

(12) Dinitrogen tetroxide _____

(13) Nitric acid _____

(14) Cobalt (II) chloride _____

(15) Hydrosulfuric acid _____

(16) Carbon tetrachloride _____

8. Give the formulas for the following salts, which contain more than one positive ion.

(1) Sodium hydrogen carbonate

(2) Magnesium ammonium phosphate

(3) Sodium potassium sulfate

(4) Ammonium hydrogen sulfide

(5) Potassium hydrogen sulfate

(6) Potassium aluminum sulfate

9. With which of the following ions can the magnesium ion combine to form a compound of formula MgZ?

(1) sulfide ion   (2) cyanide ion

(3) bromide ion   (4) carbonate ion

(5) nitrate ion   (6) iron (II) ion

10. Write the correct formulas for compounds formed by matching each cation with all anions.

	Cl$^-$	S^{2-}	NO$_3{}^-$	PO$_4{}^{3-}$
Mg^{2+}				
Fe^{3+}				
H$^+$				
NH$_4{}^+$				
Sn^{4+}				

11. What is missing from or wrong with following names of compounds?

(1) carbon oxide   (2) cobalt chloride   (3) barium monosulfide tetroxide

12. For which of the names below are you able to write the chemical formula without additional information? Explain briefly.

(1) silver chloride   (2) mercury bromide   (3) zinc oxide   (4) copper sulfide

ANSWERS TO QUESTIONS

1. (1) S   (2) S^{2-} (3) Li   (4) Li^+ (5) P   (6) P^{3-}

2. Recall that electrons that are lost are written as products (chemical equations do not "minus" substances).

$$S + 2e^- \rightarrow S^{2-}; \; Li \rightarrow Li^+ + e^-; \; P + 3e^- \rightarrow P^{3-}$$

3. (1) Al^{3+}, O^{2-}, Al_2O_3 (2) $Fe^{3+}, Br^-, FeBr_3$
   (3) $NH_4^+, SO_4^{2-}, (NH_4)_2SO_4$ (4) $Mg^{2+}, PO_4^{3-}, Mg_3(PO_4)_2$
   (5) $Cu^{2+}, CN^-, Cu(CN)_2$ (6) K^+, CO_3^{2-}, K_2CO_3

4. Note that all the cation names should be followed by the word "ion". Remember, ionic compounds are generally split into only two parts – a cation and an anion.

Formula	cation symbol	name	Roman numeral?	anion symbol	name	COMPOUND NAME
KNO_3	K^+	potassium	no	NO_3^-	nitrate	potassium nitrate
$SnCl_4$	Sn^{4+}	tin	IV	Cl^-	chloride	tin (IV) chloride
Na_2SO_4	Na^+	sodium	no	SO_4^{2-}	sulfate	sodium sulfate
$K_2Cr_2O_7$	K^+	potassium	no	$Cr_2O_7^{2-}$	chromate	potassium chromate
$FeCl_3$	Fe^{3+}	iron	III	Cl^-	chloride	iron (III) chloride
CrO_3	Cr^{6+}	chromium	VI	O^{2-}	oxide	chromium (VI) oxide
NH_4NO_3	NH_4^+	ammonium	no	NO_3^-	nitrate	ammonium nitrate

5. Note that no prefix is needed if the first nonmetal is a "mono", and that the name of the second nonmetal has been modified to the "ide" ending.

Formula	prefix	name of 1st nonmetal	prefix	name of 2nd nonmetal	NAME
SO_3	—	sulfur	tri	ox**ide**	sulfur trioxide
Si_2Cl_6	di	silicon	hexa	chlor**ide**	disilicon hexachloride
IBr	—	iodine	mono	brom**ide**	iodine monobromide
BrF_5	—	bromine	penta	fluor**ide**	bromine pentafluoride

6. (HCl)   CO SnO_2 Al_2O_3 PBr_3 $PbCl_2$

   $MgCl_2$ (H_2CO_3)   NaCl SO_3 (H_2SO_4) N_2O_3

7. (1) Aluminum phosphate   (9) $MgCl_2$
   (2) Ammonium carbonate   (10) SiO_2
   (3) Sodium iodide   (11) H_2SO_4
   (4) Barium oxide   (12) N_2O_4
   (5) Sodium hydrogen carbonate   (13) HNO_3
   (6) HCl   (14) $CoCl_2$
   (7) PbO   (15) H_2S
   (8) $KMnO_4$ (16) CCl_4

8. (1) $NaHCO_3$ (2) $MgNH_4PO_4$ (3) $NaKSO_4$
   (4) NH_4HS (5) $KHSO_4$ (6) $KAl(SO_4)_2$

9. (1) sulfide ion   (4) carbonate ion

10.

$MgCl_2$	MgS	$Mg(NO_3)_2$	$Mg_3(PO_4)_2$
$FeCl_3$	Fe_2S_3	$Fe(NO_3)_3$	$FePO_4$
HCl	H_2S	HNO_3	H_3PO_4
NH_4Cl	$(NH_4)_2S$	NH_4NO_3	$(NH_4)_3PO_4$

11. (1) The second part of the name must include a prefix, even if it is only for one atom of that element. So although we know it is a C and an O in the compound, we don't know how many O's. We assume there is only one C, as "mono" is not needed for the first element.

(2) There is more than one possible cobalt ion. Is the cobalt in this compound made up of Co^{2+} or Co^{3+} ions? The appropriate Roman numeral after the cobalt is needed.

(3) Since barium is a metal, it is assumed the compound is ionic. The second half of the name should be the name of the polyatomic ion, not named as though it is a binary nonmetallic compound.

12. (1) and (3) don't need any further information. Although silver and zinc are transition metals, they form only one ion each, Ag^+ and Zn^{2+}. Therefore their ions are unambiguous. (2) and (4), mercury and copper, have more than one choice, and therefore the charge on the cation must be specified with the Roman numeral in the name of the compound.

CHAPTER SEVEN

Quantitative Composition of Compounds

SELECTED CONCEPTS REVISITED

The mole has become the center of your universe. Not quite, but close. The mole is simply a name given to a very large number, 6.02×10^{23} (also known as Avogadro's number). A dozen is 12, a gross is 144, a mole is 6.022×10^{23}.

The molar mass of a substance is the mass in grams of 1 mole of particles of that substance. That is, if you took 6.022×10^{23} atoms of gold, put it on a scale and weighed it in grams, that would be the molar mass of gold. Numerically that would be equivalent to the atomic mass of gold. So the atomic mass of gold is 197.0, therefore, 1 mole of gold weighs 197.0 g. That is, the molar mass of gold is 197.0 g/mole.

The empirical formula for a compound is the minimum whole number atomic ratio of each element in a single unit or molecule of that compound. If the relationship between the elements is given as a mass ratio, you need to take into account that atoms of each element have different masses. Therefore the masses should first be converted to moles and since moles represent a numerical quantity of atoms, the number of atoms of each element can then be compared.

COMMON PITFALLS TO AVOID

Remember that percent composition refers to the percent by mass and do not forget to take into account the numbers of each type of atom present when calculating percent composition. Thus the percent composition of C in C_2H_6 is

$$\% \text{ comp} = \frac{\text{mass of C in } C_2H_6}{\text{mass of } C_2H_6} \times 100 = \frac{(2)(12)}{(2)(12) + (6)(1)} \times 100 = 80\%$$

Do not forget to double-check your answer to see if it makes sense. If you have difficulty realizing whether you should multiply or divide by the molar mass to convert from grams to moles, do a quick qualitative check. For example, you want to convert 135 g of unknown A to moles of A and the molar mass of A is 154 g. Without even doing the actual calculation, you can see that one mole weighs 154 g so if you have less than 154 g you should have less than one mole. (Therefore you would need to divide the 135g by the molar mass of 154 g/mol.)

When taking masses from the periodic chart, be consistent about the number of significant digits used. Rarely should the mass from the periodic chart limit the significant figures in the calculation. Also, do not round off to the nearest whole number and then simply add .0 to get extra significant digits. For example, if asked how many moles are found in 2.49 g of Ni, the molar mass for Ni from the periodic chart should be 58.69 g/mole or 58.7 g/mole, NOT 59 g/mol (too few sig. figs.), nor 59.0 g/mil, (incorrect rounding off).

RECAP SECTION

Chapter 7 puts you to work using the periodic table and chemical formulas to try out your math skills on some basic chemical problem solving. The concepts of the mole, molar mass, and Avogadro's number to calculate the molar mass of a compound, the percent composition of a compound from its chemical formula and from experimental data, and then determining its empirical formula and molecular formula are all covered in this chapter. Prior to the late 1890's, many chemists spent their careers working on problems of the elemental composition of various compounds; the techniques and mathematical steps that they used are all related to the problems found in this chapter. Even today, the simple mathematical steps encountered in this chapter are routinely used by chemists just about every time they carry out a chemical reaction. These calculations will be built upon in later chapters so it essential to grasp these fundamentals now.

1. Calculate the molar mass for the following compounds with the aid of the short list of atomic masses.

		Atomic Masses (g/mol)	
(1)	CH_2Cl_2	Na	22.99
(2)	$Ca(OH)_2$	C	12.01
(3)	NH_4Cl	O	16.00
(4)	NaClO	Ca	40.08
(5)	$CaCO_3$	Cl	35.45
		N	14.01
		H	1.008

    Do your calculations here.

2. Using the molar masses that were calculated for the previous question, find the number of moles present for the following masses of chemical. Double-check your answer by making sure the units cancel out properly and by simple math logic to see if it makes sense (is the mass given more or less than the mass of 1 mole).

    (1) 77 g CH_2Cl_2
         Number of moles = _____

    (2) 31 g of $Ca(OH)_2$
         Number of moles = _____

    (3) 89 g NH_4Cl
         Number of moles = _____

    Do your calculations here.

3. Using the molar masses that were calculated for the first question, find the number of grams represented by the following number of moles of chemical. Again, double-check your answer logically.

    (1) 1.5 moles of NaClO _____ g

    (2) 0.67 mole of $CaCO_3$ _____ g

    Do your calculations here.

4. How many grams of gold (Au) contains 2.98×10^{24} atoms?

5. Using the partial list of atomic masses, calculate the number of moles represented by a certain mass of an element.

        Na   22.99 g/mol
        Ag   107.9 g/mol
        S   32.07 g/mol

(1)   How many moles are represented by 0.460 g of Na atoms?

Do your calculations here.

(2)   How many moles are represented by 18.9 g of Ag?

Do your calculations here.

(3)   How many moles are represented by 79.5 g of S?

Do your calculations here.

6. Using the following list of molar masses, calculate the elemental percent composition for the four compounds listed.

Molar masses   N = 14.01 g/mol,   O = 16.00 g/mol,   Al = 26.98 g/mol
                         C = 12.01 g/mol,   Mn = 54.94 g/mol,   K = 39.10 g/mol

(1) N_2O (2) $Al_2(CO_3)_3$ (3) CO_2 (4) $KMnO_4$

Do your calculations here.

(1) N_2O

(2) $Al_2(CO_3)_3$

(3) CO_2

(4) $KMnO_4$

7. Fill in the blank space or circle the appropriate response.

The simplest formula of a compound, or the (1) _____ formula, tells us the (2) smallest/largest ratio of atoms present in a compound. The ratio is usually a small (3) _____ number ratio. It is possible for two compounds to have the same empirical formula, but different (4) _____ formulas. The true formula for a compound, or (5) _____ formula, represents the actual number of (6) _____ of each element found in one molecule of a compound. The mass of all the (7) _____ in a molecular formula is the compound's (8) _____.

8. A certain compound is found to contain 1.45 g of Na, 2.05 g of S, and 1.5 g of O. With the aid of the list of molar masses, calculate the empirical formula.

Na   Molar mass   22.99 g/mol
S   Molar mass   32.07 g/mol
O   Molar mass   16.00 g/mol
                                Formula_____

Do your calculations here.

9. From the percent composition data given for each compound, calculate the empirical formula.

(1)   86.6% Pb   Molar mass Pb = 207.2 g/mol
        13.4% S   Molar mass S = 32.07 g/mol
                        Empirical formula_____

Do your calculations here.

(2)   69.59% Ba   Molar mass Ba = 137.3 g/mol
        6.08% C   Molar mass C = 12.01 g/mol
        24.33% O   Molar mass O = 16.00 g/mol
                        Empirical formula._____

Do your calculations here.

10. The molar mass of an unknown sugar substance is experimentally found to be 180 g/mol. The percent composition data are determined to be 40.% carbon, 7.0% hydrogen, and 53% oxygen. Calculate first the empirical formula and then the molecular formula.

C   Molar mass   12.01 g/mol
H   Molar mass   1.008 g/mol
O   Molar mass   16.00 g/mol
                                Empirical_____
                                Molecular_____

Do your calculations here.

11. A compound of bromine and iodine is formed by direct reaction between the two elements. It is found that 5.40 g of bromine react with 8.58 g of iodine. What is the empirical formula for the compound and if the molar mass is 206.8 g/mol, what is the molecular formula? What is the percent composition of the compound?

    Do your calculations here.

12. Calculate the number of grams and the number of atoms represented by the following quantities of two elements.

    (1) 2.25 mol of Ni   (2) 0.036 mol of Ag
         Molar mass of Ni $=$ 58.69 g/mol   Molar mass of Ag $=$ 107.9 g/mol

    Do your calculations here.

13. How many grams of Fe contain the same number of atoms as 154 g of arsenic?

    Do your calculations here.

Challenge Problems

14. A compound known as cadaverine (1,5-pentane diamine) is a ptomaine formed by the action of bacteria on meat and fish. Analysis shows that the elemental composition is C 58.8%, H 13.8% and N 27.4%. Determine the empirical formula and the molecular formula. ("Pentane" means 5 carbon atoms.)

    Do your calculations here.

15. The population of the world is estimated to be 7.016×10^9 (7016 million) people by May 2012. What would result in each person having more money, giving everyone 99990000 pennies ($999900) or distributing one mole of pennies equally among the entire population?

ANSWERS TO QUESTIONS AND SOLUTIONS TO PROBLEMS

1. Sum of the average masses of each atom in the molecule.

    (1) 12.01 g/mol $+$ (2 $\times$ 1.008) g/mol $+$ (2 $\times$ 35.45) g/mol $=$ 84.93 g/mol
    (2) 40.08 g/mol $+$ (2 $\times$ 16.00) g/mol $+$ (2$\times$ 1.008) g/mol $=$ 74.096 $=$ 74.10 g/mol
    (3) 14.01 g/mol $+$ (4 $\times$ 1.008) g/mol $+$ 35.45 g/mol $=$ 53.49 g/mol
    (4) 22.99 g/mol $+$ 35.45 g/mol $+$ 16.00 g/mol $=$ 74.44 g/mol
    (5) 40.08 g/mol $+$ 12.01 g/mol $+$ (3 $\times$ 16.00) g/mol $=$ 100.09 g/mol

2.   (1)   0.91 mol

molar mass of $CH_2Cl_2 = 84.93$ g/mol
The conversion factor is either

$$\frac{84.93 \text{ g}}{1 \text{ mol}} \quad \text{OR} \quad \frac{1 \text{ mol}}{84.93 \text{ g}}$$

Therefore,

$$(77 \text{ g})\left(\frac{1 \text{ mol}}{84.93 \text{ g}}\right) = 0.91 \text{ mol (2 sig figs)}$$

   (2)   0.42 mol
molar mass of $Ca(OH)_2 = 74.10$ g/mol
The conversion factor is either

$$\frac{74.10 \text{ g}}{1 \text{ mol}} \quad \text{OR} \quad \frac{1 \text{ mol}}{74.10 \text{ g}}$$

Therefore,

$$(31 \text{ g})\left(\frac{1 \text{ mol}}{74.10 \text{ g}}\right) = 0.42 \text{ mol}$$

   (3)   1.7 mol
molar mass of $NH_4Cl = 53.49$ g/mol
The conversion factor is either

$$\frac{53.49 \text{ g}}{1 \text{ mol}} \quad \text{OR} \quad \frac{1 \text{ mol}}{53.49 \text{ g}}$$

Therefore,

$$(89 \text{ g})\left(\frac{1 \text{ mol}}{53.49 \text{ g}}\right) = 1.7 \text{ mol}$$

3.   (1)   1.5 moles of NaClO will equal

$$(1.5 \text{ mol})\left(\frac{74.44 \text{ g}}{1 \text{ mol}}\right) = 1.1 \times 10^2 \text{ g}$$

   (2)   0.67 mole of $CaCO_3$ will equal

$$(0.67 \text{ mol})\left(\frac{100.09 \text{g}}{1 \text{mol}}\right) = 67 \text{ g}$$

4.   (1)   Info given or known: 2.98×10^{24} atoms of gold
                          avg mass of gold (from Periodic Chart) = 197.0 g/mol
                          6.022×10^{24} atoms per mole
Solution map: atoms Au→moles Au→g Au

Conversion factors:

$$\frac{1 \text{ mol}}{197.0 \text{ g}} \quad \text{OR} \quad \frac{197.0 \text{ g}}{1 \text{ mol}} ; \quad \frac{6.022 \times 10^{23} \text{ atoms}}{1 \text{ mol}} \quad \text{OR} \quad \frac{1 \text{ mol}}{6.022 \times 10^{23} \text{ atoms}}$$

Therefore:

$$\left(2.98 \times 10^{24} \text{ atoms}\right)\left(\frac{1 \text{ mol}}{6.022 \times 10^{23} \text{ atoms}}\right)\left(\frac{197.0 \text{ g}}{1 \text{ mol}}\right) = 975 \text{ g (3 sig figs)}$$

5.   (1)   0.0200 moles of Na

You are asked how many moles are contained in 0.460 g of Na atoms. One mole of any element is equal to the atomic mass expressed in grams. For Na this amount is 22.99 g. What we have said is that 22.99 g of Na equals 1.0 mol. We have 0.460 g, which is less than 1.0 mole. Arranging the problem so that the units will cancel out properly and give us an answer that is less than one, we have

$$(0.460 \text{ g})\left(\frac{1 \text{ mol}}{22.99 \text{ g}}\right) = 0.0200 \text{ mol}$$

(2) 0.175 mole of Ag

1 mole of Ag $= 107.9$ g

Therefore,

$$(18.9 \text{ g}) \left(\frac{1 \text{ mol}}{107.9 \text{ g}} \right) = 0.175 \text{ mol}$$

(3) 2.48 moles of S

The problem is solved just as the others have been.

$$(79.5 \text{ g}) \left(\frac{1 \text{ mol}}{32.07 \text{ g}} \right) = 2.48 \text{ mol S}$$

6. (1) N_2O

molar mass $= (2 \times 14.01 \text{ g/mol}) + 16.00 \text{ g/mol} = 44.02 \dfrac{\text{g}}{\text{mol}}$

Therefore,

$$\%N = \frac{2 \times 14.01 \text{ g/mol}}{44.02 \text{ g/mol}} \times 100 = 63.65\%$$

$$\%O = \frac{16.00 \text{ g/mol}}{44.02 \text{ g/mol}} \times 100 = 36.35\%$$

(2) Al_2O_3

molar mass $= (2 \times 26.98 \text{ g/mol}) + (3 \times 16.00 \text{ g/mol}) = 102.0 \dfrac{\text{g}}{\text{mol}}$

Therefore,

$$\%Al = \frac{2 \times 26.98 \text{ g/mol}}{102.0 \text{ g/mol}} \times 100 = 52.90\%$$

$$\%O = \frac{3 \times 16.00 \text{ g/mol}}{102.0 \text{ g/mol}} \times 100 = 47.06\%$$

(3) K_2CO_3

molar mass $= (2 \times 39.10 \text{ g/mol}) + 12.01 \text{ g/mol} + (3 \times 16.00 \text{ g/mol}) = 138.2 \dfrac{\text{g}}{\text{mol}}$

Therefore,

$$\%K = \frac{2 \times 39.10 \text{ g/mol}}{138.2 \text{ g/mol}} \times 100 = 56.58\%$$

$$\%C = \frac{12.01 \text{ g/mol}}{138.2 \text{ g/mol}} \times 100 = 8.690\%$$

$$\%O = \frac{3 \times 16.00 \text{ g/mol}}{138.2 \text{ g/mol}} \times 100 = 34.73\%$$

(4) $KMnO_4$

molar mass $= 39.10 \text{ g/mol} + 54.94 \text{ g/mol} + (4 \times 16.00 \text{ g/mol}) = 158.0 \dfrac{\text{g}}{\text{mol}}$

Therefore,

$$\%K = \frac{39.10 \text{ g}}{158.0 \text{ g}} \times 100 = 24.75\%$$

$$\%Mn = \frac{54.94 \text{ g}}{158.0 \text{ g}} \times 100 = 34.77\%$$

$$\%O = \frac{4 \times 16.00 \text{ g}}{158.0 \text{ g}} \times 100 = 40.51\%$$

7. (1) empirical   (2) smallest   (3) whole   (4) molecular
    (5) molecular   (6) atoms   (7) atoms   (8) molar mass

8. $Na_2S_2O_3$

    Solution Map: g element⇒moles element⇒mol ratios

    We must determine the number of moles of each element present and then find the smallest whole-number ratio.

    Smallest Ratio

    Na $(1.45 \text{ g})\left(\dfrac{1 \text{ mol}}{22.99 \text{ g}}\right) = 0.0631 \text{ mole}$ $\dfrac{0.0631}{0.0631} = 1$

    S $(2.05 \text{ g})\left(\dfrac{1 \text{ mol}}{32.07 \text{ g}}\right) = 0.0639 \text{ mole}$ $\dfrac{0.0639}{0.0631} = 1$

    O $(1.5 \text{ g})\left(\dfrac{1 \text{ mol}}{16.00 \text{ g}}\right) = 0.094 \text{ mole}$ $\dfrac{0.094}{0.0631} = 1.5$

    So, 1 : 1 : 1.5 becomes 2 : 2 : 3

    The empirical formula is $Na_2S_2O_3$

9. (1) PbS   (2) $BaCO_3$

    (1) Taking 100 g of the compound composed of Pb and S, we would have 87 g Pb and 13 g S. We need to determine the number of moles of Pb and S present in the 100 g.

    Therefore,

    Pb $(87 \text{ g})\left(\dfrac{1 \text{ mol}}{207.2 \text{ g}}\right) = 0.42 \text{ mol}$

    S $(13 \text{ g})\left(\dfrac{1 \text{ mol}}{32.07 \text{ g}}\right) = 0.41 \text{ mol}$

    The ratio of Pb to S is 0.42 to 0.41 or 1:1. The formula is PbS.

    (2) Likewise, in a 100 g sample of the Ba, C, O compound, we would have 69.59 g of Ba, 6.08 g of C, and 24.33 g of O. The number of moles of each element would be:

    Ba $(69.59 \text{ g})\left(\dfrac{1 \text{ mol}}{137.3 \text{ g}}\right) = 0.5068 \text{ mol}$

    C $(6.08 \text{ g})\left(\dfrac{1 \text{ mol}}{12.01 \text{ g}}\right) = 0.506 \text{ mol}$

    O $(24.33 \text{ g})\left(\dfrac{1 \text{ mol}}{16.00 \text{ g}}\right) = 1.521 \text{ mol}$

    To eliminate the decimals from our ratio we must divide each of the numbers by the smallest number.

    $Ba = \dfrac{0.5068}{0.506} = 1$ $O = \dfrac{1.521}{0.506} = 3$

    $C = \dfrac{0.506}{0.506} = 1$

    When the ratio is expressed as small whole numbers, the correct empirical formula becomes $BaCO_3$.

10. Empirical formula: CH_2O Molecular formula: $C_6H_{12}O_6$

    The first step is to find the empirical formula from the number of moles of each element for a hypothetical 100 g of compound.

Smallest Ratio

$$C \quad (40. \text{ g})\left(\frac{1 \text{ mol}}{12.01 \text{ g}}\right) = 3.3 \text{ moles} \qquad \frac{3.3}{3.3} = 1$$

$$H \quad (7.0 \text{ g})\left(\frac{1 \text{ mol}}{1.008 \text{ g}}\right) = 6.9 \text{ moles} \qquad \frac{6.9}{3.3} = 2.1$$

$$O \quad (53 \text{ g})\left(\frac{1 \text{ mol}}{16.00 \text{ g}}\right) = 3.3 \text{ moles} \qquad \frac{3.3}{3.3} = 1$$

The rounded-off whole-number ratio would be $(GH_2O)_n$ and the empirical formula would therefore be CH_2O.

The molecular formula is calculated from the total atomic masses in the empirical formula and the molar mass. The total mass of the empirical formula is $12.01 + 2.016 + 16.00 = 30.03$. Next, determine the ratio between the given molar mass and the empirical formula mass.

$$\frac{180 \text{ g/mol}}{30.03 \text{ g/mol}} = 5.99$$

Therefore, we need to multiply the empirical formula by 6 to obtain the molecular formula.

$$6 \times CH_2O \text{ would be } C_6H_{12}O_6$$

11. Empirical formula and molecular formula are the same, BrI. The percent composition is 38.6% Br_2 and 61.3% I_2.

$$\text{Number of moles of } Br_2 = \frac{5.40 \text{ g}}{79.90 \text{ g/mol}} = 0.0676 \text{ mol}$$

$$\text{Number of moles of } I_2 = \frac{8.58 \text{ g}}{126.9 \text{ g/mol}} = 0.0676 \text{ mol}$$

The number of moles of each element are in a ratio of 1:1, so the empirical formula is BrI. The mass of one mole of BrI is 206.8 g, which matches the value given in the problem, so the molecular formula is also BrI. Percent composition is calculated as follows:

$$\%Br = \frac{\text{mass } Br_2}{\text{total mass}} = \frac{5.40 \text{ g}}{(5.40 + 8.58) \text{ g}} \times 100 = 38.6\%$$

Likewise

$$\%I_2 = \frac{8.58 \text{ g}}{13.98 \text{ g}} \times 100 = 61.4\%$$

12. (1) 132 g of Ni and 1.35×10^{23} atoms of Ni

   Nickel has 58.69 g in 1 mole. We have 2.25 moles, which will be more than 58.69 g. Arranging the problem so that the units cancel out, we have:

   $$(2.25 \text{ mol})\left(58.69 \frac{\text{g}}{\text{mol}}\right) = 132 \text{ g of Ni}$$

   To find the number of atoms in a certain number of moles, we simply multiply the number of moles by Avogadro's number.

   $$(2.25 \text{ mol})\left(6.022 \times 10^{23} \frac{\text{atoms}}{\text{mol}}\right) = 1.35 \times 10^{24} \text{ atoms Ni}$$

   (2) 3.9 g of Ag and 2.2×10^{22} atoms Ag

   $$(0.036 \text{ mol})\left(107.9 \frac{\text{g}}{\text{mol}}\right) = 3.9 \text{ g of Ag}$$

   $$(0.036 \text{ mol})\left(6.022 \times 10^{23} \frac{\text{atoms}}{\text{mol}}\right) = 2.2 \times 10^{22} \text{ atoms Ca}$$

13. 115 g Fe. We can start by finding out how many moles of arsenic are represented by 154 g.

$$\frac{154 \text{ g}}{74.92 \text{ g/mol}} = 2.06 \text{ mol As}$$

2.06 moles of As contains the same number of atoms as 2.06 moles of Fe. To convert this value into grams of Fe, we need only to multiply the molar mass by the number of moles.

$$(2.06 \text{ mol}) \left(55.85 \frac{\text{g Fe}}{\text{mol}} \right) = 115 \text{ g Fe (3 significant figures)}$$

14. The empirical and molecular formula are both $C_5H_{14}N_2$, since we know from the problem that the compound contains 5 C atoms.

$$C \quad \frac{58.5 \text{ g}}{12.01 \text{ g/mol}} = 4.90 \text{ mol}$$

$$N \quad \frac{27.4 \text{ g}}{14.01 \text{ g/mol}} = 1.96 \text{ mol}$$

$$H \quad \frac{13.8 \text{ g}}{1.008 \text{ g/mol}} = 13.7 \text{ mol}$$

15. Distributing one mole of pennies among the entire population would give each person more money.

The problem can be solved by comparing either the moles of pennies distributed in both cases, or the amount of money each person receives.

Comparing moles of pennies distributed:

if everyone were give $999900, then 99990000 pennies would be distributed per person, so total pennies given:

population . . . 7.016×10^9 persons
99990000 pennies per person
1 mol pennies = 6.022×10^{23} pennies
solution map: # people ⇒ # pennies ⇒ mol pennies

$$7.016 \times 10^9 \text{ persons} \times \frac{99990000 \text{ pennies}}{1 \text{ person}} \times \frac{1 \text{ mol}}{6.022 \times 10^{23} \text{ pennies}}$$
$$= 1.165 \times 10^{-6} \text{ moles of pennies distributed}$$

Less than 1 mole of pennies is distributed in this scenario.

OR . . . comparing money each person receives:

population . . . 7.016×10^9 persons
1 dollar = 100 pennies
1 mol pennies = 6.022×10^{23} pennies
solution map: 1 mol pennies ⇒ dollars ⇒ dollars per peson

$$1 \text{ mol of pennies} \times \frac{6.022 \times 10^{23} \text{ pennies}}{1 \text{ mol}} \times \frac{1 \text{ dollar}}{100 \text{ pennies}} \times \frac{1}{7.016 \times 10^9 \text{ persons}} = \$ 8.583 \times 10^{11}$$

In other words, 1 mol of pennies distributed equally would give over 850 billion dollars to each person!

CHAPTER EIGHT

Chemical Equations

SELECTED CONCEPTS REVISITED

Understanding and balancing chemical equations is a key aspect of any chemistry course. This is how chemists communicate reactions to each other, regardless of language. When balancing an equation, you cannot change the molecular formula of any substance. Only the coefficients of the substances may change. Therefore be certain to establish the molecular formula for all the reactants and products <u>before</u> you begin to balance the reaction. You may only change the amount of each substance present by changing the coefficients of the compounds.

When you place a coefficient in front of a formula it multiples every atom in the formula by that amount. Therefore when you are counting atoms, multiply any subscripts by the coefficient to get the number of that type of atom present. Bear in mind that the subscript after a set of atoms in parentheses applies to each atom in the parentheses.

For example, $3 \ Al_2(SO_4)_3$ has:
$(3)(2) \ Al = 6 \ Al$
$(3)(1)(3) \ S = 9 \ S$
$(3)(4)(3) \ O = 36 \ O$

In many cases where there are identical polyatomic ions on both sides of the equation, you can simply count the polyatomics as a unit and balance them instead of breaking them down into individual atoms. This shortcut is most useful when balancing double-displacement reactions.

Balancing chemical reactions is largely trial and error. You will only get proficient at balancing equations by practicing. However, this is one of the easiest types of problems to double-check to see if you have the right answer – either the atoms are balanced when you are finished or they are not. Always check your balanced equations.

You have also been introduced to the activity series for metals. Notice the general trend of the more reactive the metal is, the easier it is to replace H in water or acids. It is easier to replace H from an acid than from water and to replace H from hot water than from cold water. So only the most reactive metals will react with cold water.

In a chemical equation, we write "+ heat", not "− heat". For example, an exothermic reaction would be written as

$$A \rightarrow B + heat \qquad NOT \qquad A - heat \rightarrow B$$

(Total mass of reactants) = (Total mass of products)   *Law of Conservation of Mass*

COMMON PITFALLS TO AVOID

Do not try to do more than one step at a time when balancing equations. Make sure you have the right formula for the reactants and products and then start to balance your equation changing only the coefficients. It may help you to think of it as two problems – (1) write the molecular formula for each reactant and product (this involves putting the right subscripts on the atoms and (2) balance the equation (this involves changing the coefficients).

When you write the molecular formula for the products of a double displacement reaction do not forget you must balance the charges to determine the formula. You will be putting together a cation and an anion but the subscripts of the cation and the anion (i.e. the subscripts in the formula) are not necessarily the same as the subscripts associated with them in the reactants. Once you know the formula and charge of each ion, ignore the reactants and determine the formula for the compound based on the relative charges of the ions. If you are comfortable with naming ionic compounds you can often double-check the formula of double-displacement reaction by determining the name of the compound first and then writing the formula.

For example, what are the formulas for the products when $AgNO_3$ reacts with $MgCl_2$?

Here is one possible approach: you know the cation from the first compound will combine with the anion from the second compound so name the respective ions (silver ion and chloride ion) so one product is silver chloride. Then simply write the formula for silver chloride, $AgCl$ because you are combining Ag^+ and Cl^- (not $AgCl_2$ which incorrectly assumes that the subscript associated with the ion in the reactant is the same as the subscript in the product).

RECAP SECTION

Chapter 8 describes the information present in a chemical equation. You have learned how to write and balance chemical equations, and how to classify many of these equations as combination, single or double displacement, or decomposition reactions. Energy profiles or heat written as a reactant or product indicate if a reaction is exothermic or endothermic. The overall heat of the reaction and the activation energy may also be found on the energy profiles. Nomenclature and balancing equations are part of the language necessary to convey chemical processes; refer to this chapter and Chapter 6 should you need refreshing on these topics.

SELF-EVALUATION SECTION

1. Match the symbols used in chemical equations with the corresponding descriptive statements.

Symbols

$\rightarrow$ (s)

$+$ (l)

$\rightleftarrows$ (g)

(aq) (Δ)

(1) Gas (written after substance) _____

(2) Reversible reaction; equilibrium between reactants and products _____

(3) Heat _____

(4) Added to _____

(5) Liquid (written after substance) _____

(6) Aqueous solution (substance dissolved in water) _____

(7) Yields; produces (points to products) _____

(8) Solid (written after substance) _____

2. Consider the equation $2\,Al(OH)_3 + 3\,H_2SO_4 \rightarrow Al_2(SO_4)_3 + 6\,H_2O$

Label the following parts of the equation: subscripts, reactants, coefficients, products.

3. Balance the following equations or translate the word equations in formulas and then balance them. The first two word-to-formula equations emphasize the process involved. You should apply the same procedure whenever you are asked to write a balanced formula equation.

(1) $H_2O_2 \rightarrow H_2O + O_2$

(2) $NH_4NO_2 \rightarrow N_2 + H_2O$

(3)   Potassium nitrate $\rightarrow$ Potassium nitrite + Oxygen
        formula: _____ $\rightarrow$ _____ + _____
        balanced equation: (note-only coefficients change)

(4)   Calcium oxide + Hydrochloric acid $\rightarrow$ Calcium chloride + Water
        formula: _____ + _____ $\rightarrow$ _____ + _____
        balanced equation: (note-only coefficients change)

(5) Bromine + Hydrogen sulfide $\rightarrow$ Hydrogen bromide + Sulfur

(6) Copper metal + Sulfuric acid $\rightarrow$ Copper (II) sulfate + Water + Sulfur dioxide

(7) $K + H_2O \rightarrow KOH + H_2$

(8) $N_2 + H_2 \rightarrow NH_3$

(9) Sodium hydrogen carbonate + Sulfuric acid $\rightarrow$ Sodium sulfate + Water + Carbon dioxide

(10) $Fe + H_2O \rightarrow Fe_3O_4 + H_2$

(11) Zinc sulfide + Oxygen $\rightarrow$ Zinc oxide + Sulfur dioxide

(12) $C_6H_{14} + O_2 \rightarrow CO_2 + H_2O$

4. Identify the following reactions as exothermic (exo) or endothermic (endo).

    (1) $N_2(g) + O_2(g) + 181 \text{ kJ} \rightarrow 2 NO(g)$ _____

    (2) $C(s) + O_2(g) \rightarrow CO_2(g) + 94.0 \text{ kcal}$ _____

    (3) $C_3H_8(g) + 5 O_2(g) \rightarrow 3 CO_2(g) + 4 H_2O(g) + 2200 \text{ kJ}$ _____

5. Identify the following reactions as combinations (C), decomposition (D), single displacement (SD), or double displacement (DD).

    (1) $K_2CrO_4 + Pb(NO_3)_2 \rightarrow 2 KNO_3 + PbCrO_4$ _____

    (2) $H_2 + Cl_2 \rightarrow 2 HCl$ _____

    (3) $Mg + NiCl_2 \rightarrow Ni + MgCl_2$ _____

    (4) $2 HgO \rightarrow 2 Hg + O_2$ _____

    (5) $Mg + H_2SO_4 \rightarrow MgSO_4 + H_2$ _____

    (6) $CaO + H_2O \rightarrow Ca(OH)_2$ _____

    (7) $NaCl + H_2SO_4 \rightarrow NaHSO_4 + HCl$ _____

    (8) $2 KClO_3 \rightarrow 2KCl + 3 O_2$ _____

6. Interpret the odd-numbered reactions of question 4 in terms of number of moles of reactants and products involved. Write your answers below.

    (1)

    (3)

Challenge Problems

7. Identify each of following reactions as combination, decomposition, single displacement or double displacement. Complete and balance each one, converting names into formula when necessary.

    (1) aluminum hydroxide + sulfuric acid $\rightarrow$

    (2) $Na + H_2O \rightarrow$

    (3) $HCl + K_2CO_3 \rightarrow$

    (4) $H_2 + N_2 \xrightarrow{\Delta}$

    (5) calcium carbonate $\xrightarrow{\Delta}$

    (6) $Zn + Pb(NO_3)_2 \rightarrow$

    (7) $Al_2(SO_4)_3 + NH_4OH \rightarrow$

    (8) Calcium oxide + water $\rightarrow$

    (9) $Cl_2 + KBr \rightarrow$

(10) $NH_3 + HCl$

(11) $Ba(NO_3)_2 + Na_2SO_4 \rightarrow$

(12) $K + O_2 \rightarrow$

(13) chlorine gas + sodium bromide $\rightarrow$

(14) nitric acid + sodium hydroxide $\rightarrow$

(15) Hydrochloric acid + sodium carbonate $\rightarrow$

(16) $MgCl_2 + AgNO_3 \rightarrow$

(17) $Na_2O + H_2SO_4 \rightarrow$

(18) $Mg + O_2 \xrightarrow{\Delta}$

(19) $NaI + Cl_2 \rightarrow$

(20) silver nitrate + sodium chloride $\rightarrow$

8. Will a reaction occur when the following are mixed? If so, write a balanced equation. Use the short activity series in the chapter.

(1) $Al(s) + NaBr(aq)$ (4) $Zn(s) + CuCl_2(aq)$ (7) $Fe(s) + H_2O(l)$
(2) $Cu(s) + HCl(aq)$ (5) $Mg(s) + NiCl_2(aq)$ (8) $Ni(s) + H_2O(g)$
(3) $K(s) + H_2O$ (6) $Sn(s) + H_2O(g)$

9. An aqueous reaction with this energy profile was carried out in a beaker. If you touched the beaker, would the glass be hot or cold? Why? Would heat be written as a reactant or product for this reaction?

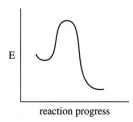

E

reaction progress

10.

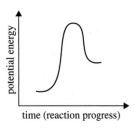

potential energy

time (reaction progress)

(1) Does the potential energy diagram show an exothermic or an endothermic reaction?

(2) Would the energy term be written as a reactant or product in the chemical equation?

(3) On the graph, show clearly the activation energy for the reaction.

1. (1) *(g)*   (2) $\leftrightarrows$ (3) Δ (4) $+$
    (5) *(l)*   (6) *(aq)*   (7) $\rightarrow$ (8) *(s)*

2. $2\,Al(OH)_3$ $+$ $3\,H_2SO_4$ $\rightarrow$ $Al_2(SO_4)_3$ $+$ $6\,H_2O$

    Coefficients   2   3   1   6

    Reactants $Al(OH)_3$ H_2SO_4

    Subscripts (shown in bold) $2\,Al(OH)_\mathbf{3}$ $+$ $3\,H_\mathbf{2}SO_\mathbf{4}$ $\rightarrow$ $Al_2(SO_4)_3 + 6\,H_2O$

    Products $Al_2(SO_4)_3$ H_2O

3. (1) $2\,H_2O_2 \rightarrow 2\,H_2O + O_2$

    (2) $NH_4NO_2 \rightarrow N_2 + 2\,H_2O$

    (3) formula: $KNO_3 \rightarrow KNO_2 + O_2$
    balanced equation: $2\,KNO_3 \rightarrow 2\,KNO_2 + O_2$

    (4) formula: $CaO + HCl \rightarrow CaCl_2 + H_2O$
    balanced equation: $CaO + HCl \rightarrow CaCl_2 + H_2O$

    (5) $Br_2 + H_2S \rightarrow 2\,HBr + S$

    (6) $Cu + 2\,H_2SO_4 \rightarrow CuSO_4 + 2\,H_2O + SO_2$

    (7) $2\,K + 2\,H_2O \rightarrow 2\,KOH + H_2$

    (8) $N_2 + 3\,H_2 \rightarrow 2\,NH_3$

    (9) $2\,NaHCO_3 + H_2SO_4 \rightarrow Na_2SO_4 + 2\,H_2O + 2\,CO_2$

    (10) $3\,Fe + 4\,H_2O \rightarrow Fe_3O_4 + 4\,H_2$

    (11) $2\,ZnS + 3\,O_2 \rightarrow 2\,ZnO + 2\,SO_2$

    (12) $2\,C_6H_{14} + 19\,O_2 \rightarrow 12\,CO_2 + 14\,H_2O$

4. When heat is written as a reactant, the reaction is endothermic.
    (1) endo   (2) exo   (3) exo

5. (1) DD   (2) C   (3) SD   (4) D
    (5) 5D   (6) C   (7) DD   (8) D

6. (1)   Reactants   1 mole N_2 and 1 mole O_2 (and 181 kJ of energy)
            Products   2 moles NO

    (3)   Reactants   1 mole C_3H_8 and 5 moles O_2
            Products   3 moles CO_2 and 4 moles H_2O (and 2200 kJ of energy)

7. (1) double displacement $2\,Al(OH)_3 + 3\,H_2SO_4 \rightarrow Al_2(SO_4)_3 + 6\,H_2O$

    (2) single displacement $2\,Na + 2\,H_2O \rightarrow 2\,NaOH + H_2(g)$

    (3) double displacement $2\,HCl + K_2CO_3 \rightarrow 2\,KCl + H_2O + CO_2(g)$

    (4) combination $3\,H_2 + N_2 \xrightarrow{\Delta} 2\,NH_3$

    (5) decomposition $CaCO_3 \xrightarrow{\Delta} CaO + CO_2(g)$

    (6) single displacement $Zn + Pb(NO_3)_2 \rightarrow Zn(NO_3)_2 + Pb$

    (7) double displacement $Al_2(SO_4)_3 + 6\,NH_4OH \rightarrow 3\,(NH_4)_2SO_4 + 2\,Al(OH)_3$

    (8) combination $CaO + H_2O \rightarrow Ca(OH)_2$

    (9) single displacement $Cl_2 + 2\,KBr \rightarrow Br_2 + 2\,KCl$

    (10) combination $NH_3 + HCl \rightarrow NH_4Cl$

    (11) double displacement $Ba(NO_3)_2 + Na_2SO_4 \rightarrow BaSO_4 + 2\,NaNO_3$

    (12) combination $4K + O_2 \rightarrow 2\,K_2O$

    (13) single displacement $Cl_2 + 2\,NaBr \rightarrow Br_2 + 2\,NaCl$

    (14) double displacement $HNO_3 + NaOH \rightarrow NaNO_3 + H_2O$

    (15) double displacement $2\,HCl + Na_2CO_3 \rightarrow 2\,NaCl + H_2O + CO_2$

(16)	double displacement	$MgCl_2 + 2\,AgNO_3 \rightarrow 2\,AgCl + Mg(NO_3)_2$
(17)	double displacement	$Na_2O + H_2SO_4 \rightarrow Na_2SO_4 + H_2O$
(18)	combination	$2\,Mg + O_2 \xrightarrow{\Delta} 2\,MgO$
(19)	single displacement	$2\,NaI + Cl_2 \rightarrow 2\,NaCl + I_2$
(20)	double displacement	$AgNO_3 + NaCl \rightarrow AgCl + NaNO_3$

8. (1) No

   (2) No

   (3) Yes $2\,K(s) + 2\,H_2O(l) \rightarrow 2\,KOH(aq) + H_2(g)$

   (4) Yes $Zn(s) + CuCl_2(aq) \rightarrow Cu(s) + ZnCl_2(aq)$

   (5) Yes $Mg(s) + NiCl_2(aq) \rightarrow Ni(s) + MgCl_2(aq)$

   (6) No

   (7) No (if H_2O were hot then it would have reacted)

   (8) No

9. The reaction profile shows an exothermic reaction that means heat is given off during the reaction and therefore the beaker should feel hotter after the reaction than it did before the reaction. In exothermic reactions heat can be written as a product.

10. (1) endothermic   (2) reactant (net absorption of energy)
    (3)

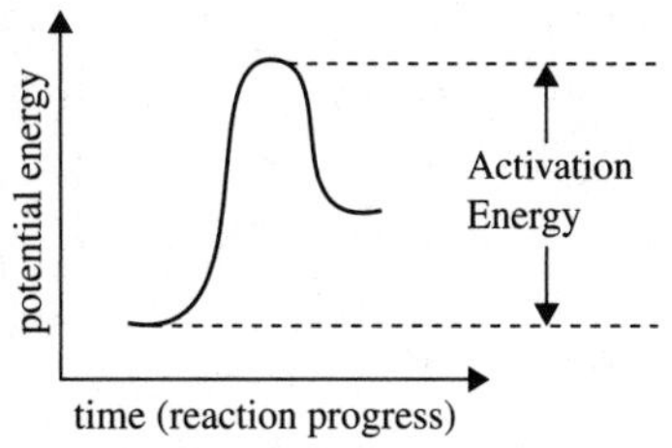

CHAPTER NINE

Calculations from Chemical Equations

SELECTED CONCEPTS REVISITED

This chapter deals with a lot of calculations and it is important that you grasp the concept and use of the mole. Two important relationships involving moles are molar mass and mole ratios.

Molar mass allows us to relate a number to a mass, that is, it relates mass to moles as the name suggests. Therefore we can translate an easily measured quantity, the mass, to the mole which allows us to compare numbers of one substance to another.

The mole ratio provides us with the numerical relationship between substances in a chemical reaction.

If you understand these relationships and where they come from, working through the problems in this chapter will be much easier. The molar mass is available from the periodic table and the mole ratio is found from a balanced chemical equation.

When carrying out calculations, ask yourself the following questions:

What am I trying to find (quantity, unit)?
What information is directly given? (e.g., a given chemical equation provides mole ratios)
What information is accessible from a periodic table or constants? (e.g., molar masses, Avogadro's number)

The limiting reactant is the reactant that runs out first based on the stoichiometry of the reaction. We need to compare moles of one reactant to moles of another using the mole ratio. Whichever reactant is used up first will limit the amount of product that can be formed. The total amount of possible product that can be formed from the amounts of reactants given is called the theoretical yield. The theoretical yield assumes that 100% of the limiting reactant is converted to product. Since this rarely occurs in practice, we generally also calculate the percent yield of the reaction which compares the actual yield to the theoretical yield.

The technique of solving mole stoichiometric problems is very important to learn. Once you have a balanced equation, it is possible to set up ratios between any two species in the equation. If the problem needs to relate grams of reactant to grams of product, there will be additional calculation steps to perform on each side of the ratio, but the ratio is the connecting link.

grams reactant A → *moles reactant A* → *moles product B* → grams product B

COMMON PITFALLS TO AVOID

You must have a balanced chemical equation in order to carry out most of the calculations or at the very least have some other way of determining the mole ratio. Do not forget to check if the equation given to you is balanced.

Use logic to avoid the tendency to use the mole ratio "upside down". Talk it out to yourself to see if it makes sense. For example, if 2 moles of product are produced for every 1 mole of reactant, then the number of moles of product should be greater than the number of moles of reactant. So if you start out with 0.68 moles of reactant, the moles of product must be greater than 0.68 so you must multiply by 2 (not divide). Dimensional analysis helps here also as long as you make sure "moles reactant" and "moles product" are used instead of just the numerical ratio.

Check with your instructor on their policy of rounding off numbers after every step of a multi-step calculation. In general, since most of these calculations involve only multiplication/division, one or two extra significant figures are carried through to the end of the calculation before the final rounding off to avoid rounding error.

The coefficients of a balanced chemical equation are irrelevant when dealing with molar mass. By definition the molar mass is the mass in grams of one mole of a substance. The number of moles of the substance being used is addressed by the mole ratio not in the molar mass. Do not use the coefficients for the molar mass.

Understand what the question is asking you to find. Too many students memorize the process: g of A to moles of A to moles of B to g of B. Then, given moles of A and asked to find g of B, students often incorrectly start the calculation using the molar mass of A (which is not needed here). The question already assumed you knew the moles of A. Read and understand the questions and understand the relationships (molar mass and mole ratio) and what you can calculate using each relationship.

Molar mass relates grams (mass) to moles for the same substance.
Mole ratio relates quantities of one substance to another substance.

RECAP SECTION

The student who masters Chapter 9 understands both the qualitative and quantitative information inherent in a balanced chemical equation. The chemical equation gives the chemicals involved. If you know the formula, you know the molar mass of the compound. The balanced equation gives the mole ratios, or stoichiometry, of the substances. Armed with this information, calculations relating either moles or masses or a combination of moles and masses of any substance within an equation should become routine. Professionals in agriculture, home economics, forestry, biology, and chemical engineering, in addition to chemists, are constantly working with stoichiometric calculations. The basic technique and approach to attacking these problems is very important to learn. Theoretical and percent yield calculations are standard practice in labs, whether academic or research.

SELF-EVALUATION SECTION

1. Fill in the blanks with the appropriate numbers that reflect the necessary moles to keep the equation balanced.

$$2\,Na + Cl_2 \rightarrow 2\,NaCl$$

(1) 2 moles of Na reacts with _____ moles of Cl_2 to produce _____ moles of NaCl.

(2) 0.50 moles of Na reacts with _____ moles of Cl_2 to produce _____ moles of NaCl.

(3) _____ moles of Na reacts with 0.040 moles of Cl_2 to produce _____ moles of NaCl.

(4) _____ moles of Na reacts with _____ moles of Cl_2 to produce 1.2 moles of NaCl.

$$Pt + 8\,HCl + 2\,HNO_3 \rightarrow H_2PtCl_2 + 2\,NOCl + 4\,H_2O$$

(5) 0.300 _____ _____ _____ _____ _____ moles

(6) _____ _____ 0.016 _____ _____ _____ moles

(7) _____ _____ _____ _____ _____ 0.100 moles

2. For the combustion of hexane as written in the equation

$$C_5H_{12} + 8\,O_2 \rightarrow 5\,CO_2 + 6\,H_2O$$

(1) what is the mole ratio of C_5H_{12} to CO_2? _____

(2) what is the mole ratio of C_5H_{12} to H_2O? _____

(3) Given 4 moles of C_5H_{12} and 30. moles of O_2 at the start of the reaction, which is the limiting reagent? _____

(4) How many moles of CO_2 can be produced from 5 moles of C_5H_{12}?

(5) If only 3.0 moles of C_5H_{12} is available for the reaction, what is the theoretical yield of H_2O in moles? _____ in grams? _____ of CO_2 in grams? _____

(6) After carrying out the above reaction with 3.0 moles of C_5H_{12}, a chemist measured an actual yield of 550 g of CO_2. What was the percent yield of CO_2?

Do your calculations here.

3. For each of the following reactions, choose the sequence of steps needed. This will help you to recognize the steps needed depending on the information given in the question.

(1) $2\,ZnS + 3\,O_2 \rightarrow 2\,ZnO + 2\,SO_2$

A chemist weighed 100. g of ZnS and burned it under the appropriate conditions. What is the theoretical yield in moles of sulfur dioxide?
(a) $g\,ZnS \rightarrow moles\,ZnS \rightarrow g\,SO_2 \rightarrow moles\,SO_2$
(b) $g\,ZnS \rightarrow moles\,ZnS \rightarrow moles\,SO_2$
(c) $g\,ZnS \rightarrow g\,SO_2 \rightarrow mol\,SO_2$
(d) $g\,ZnS \rightarrow g\,O_2 \rightarrow mol\,O_2 \rightarrow mol\,SO_2$

(2) $Na_2O + H_2O \rightarrow 2\,NaOH$

If an excess of water was mixed with 25.0 g of Na_2O under reaction conditions, what is the theoretical mass of the product?
(a) $g\,Na_2O \times 2 \rightarrow g\,NaOH$
(b) $g\,Na_2O \rightarrow mol\,NaOH \rightarrow g\,NaOH$
(c) $g\,Na_2O \rightarrow mol\,Na_2O \rightarrow g\,NaOH$
(d) $g\,Na_2O \rightarrow mol\,Na_2O \rightarrow mol\,NaOH \rightarrow g\,NaOH$

(3) A 5.00 g strip of magnesium was placed in a beaker of H_2SO_4. The magnesium strip was fully reacted and 5.97 g of $MgSO_4$ was recovered. What is the percent yield of the magnesium sulfate recovered.
(a) $g\,Mg \rightarrow theoretical\,g\,MgSO_4 \rightarrow \%\,yield$
(b) $g\,Mg \rightarrow mole\,Mg \rightarrow g\,MgSO_4\,recovered \rightarrow \%\,yield$
(c) $g\,Mg \rightarrow mol\,Mg \rightarrow mol\,MgSO_4 \rightarrow theoretical\,g\,MgSO_4 \rightarrow \%\,yield$
(d) $g\,MgSO_4\,recovered \rightarrow mol\,MgSO_4 \rightarrow mol\,Mg \rightarrow \%\,yield$

4. For the reaction in part (1) of question 3, calculate the theoretical moles of ZnO obtained if 100. g of _each_ reactant was used. (Hint, first determine the limiting reagent.)

Do your calculations here.

5. We will now use the techniques of Chapter 9 to gain further chemical equation problem-solving skill. Balance the following equation and then calculate the requested quantities.

$$PbO_2 \xrightarrow{\Delta} PbO + O_2(g)$$

(1) How many grams of O_2 can be obtained from 100. grams of PbO_2? This is the theoretical yield. Remember that we need to (a) use a balanced equation, (b) determine the number of moles of starting substance, (c) calculate the number of moles of desired substance, and (d) convert moles to grams of desired substance.

(2)   What is the percent yield if the actual yield of O_2 in the above reaction was 5.0 grams?

Do your calculations here.

6.   Very often in industrial chemical processes, one of the reactants will be present in an amount that exceeds the requirements of the balanced equation. The reactant that is not in excess will therefore limit the amount of product that is formed and is named the limiting reactant. Using the equation given, answer the questions below.

$$2\,Al(OH)_3 + 3H_2SO_4 \rightarrow Al_2(SO_4)_3 + 6\,H_2O$$

(1)   The reaction is run with excess of sulfuric acid. What is the limiting reactant?

(2)   You have 9 moles of H_2SO_4 present for the reaction. How many moles of $Al(OH)_3$ are required to react completely with this amount of H_2SO_4?

(3)   If 4 moles of $Al(OH)_3$ is the amount available, how much of the 9 moles of H_2SO_4 can be used?

(4)   Using the 4 moles of $Al(OH)_3$, how many moles of $Al_2(SO_4)_3$ and H_2O will you be able to produce?

Do your calculations here.

7.   Balance the following equation and calculate how many moles of Cu can be formed from 5.0 moles of Al and 10.0 moles of $CuSO_4$. What is the limiting reactant and how much of the excess reactant is left after the reaction?

$$\begin{array}{ccc} Al & + \quad CuSO_4 & \rightarrow Cu + Al_2(SO_4)_3 \\ 5.0\,mol & 10.0\,mol & \end{array}$$

Do your calculations here.

For questions 8 and 9 please set up your solution map before starting the calculation.

8.   In the following reaction, how many moles of ZnO can be obtained? Also determine which reactant is the limiting reactant and which reactant is in excess.

$$\begin{array}{ccc} 2\,ZnS + & 3\,O_2 & \rightarrow 2\,ZnO + 2\,SO_2 \\ 100.\,g & 100.\,g & \end{array}$$

Solution map.

Do your calculations here.

9. What is the theoretical yield of Fe_2O_3 that can be produced from 3.0 kg of Fe according to the following unbalanced equation? Balance the equation first.

$$Fe + O_2 \rightarrow Fe_2O_3$$

Solution map.

Do your calculations here.

10. Consider the reaction $2\,MX_2 + X_2 \rightarrow 2\,MX_3$.

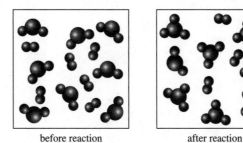

before reaction   after reaction

(1) Which reagent is the limiting reagent? (Hint, consider only the before picture.)

(2) If 0.342 moles of X_2 is used, what is the minimum amount of MX_2 needed for the reaction to go to completion?

(3) What is the actual yield of the reaction as illustrated?

Challenge Problems

11. It is possible to reclaim silver from used photographic fixer by using the active metal, powdered zinc. Zinc replaces the silver in solution, followed by conversion of the silver to silver oxide. After filtering, the silver oxide is reduced to metallic silver by carbon. The series of reactions are:

(1) $2\,AgBr + Zn \rightarrow ZnBr_2 + 2\,Ag$

(2) $4\,Ag + O_2 \rightarrow 2\,Ag_2O$

(3) $2\,Ag_2O + C \rightarrow CO_2 + 4\,Ag$

How much silver can be reclaimed from 2.000 gallons of used fixer (density 1.018 g/mL) if the silver concentration is 300 parts per million? Part per million is a general purpose concentration term that can take on a variety of units. For example 1 μg per gram, 1 μL per liter and 1 mg per kilogram are examples of 1 ppm concentration.

Do your calculations here.

12. Refer to the problem above and the last reaction. The carbon that reduces the Ag_2O to metallic Ag comes from the filter paper that was used in the filtration step following reaction (2). Assuming the filter paper to be pure cellulose with an empirical formula of $C_6H_{12}O_6$, what mass of filter paper is required to reduce the 2.31 g of silver contained in the two gallon volume of fixer?

Do your calculations here.

ANSWERS TO QUESTIONS AND SOLUTIONS TO PROBLEMS

1. (1) 1, 2   (2) 0.25, 0.50   (3) 0.080, 0.080   (4) 1.2, 0.60

(5)	0.300	2.40	0.600	0.300	0.600	1.20 moles
(6)	0.0080	0.13	0.016	0.0080	0.016	0.032 moles
(7)	0.0250	0.200	0.0500	0.0250	0.0500	0.100 moles

2. (1) 1:8   (2) 1:6   (3) O_2

(4)   moles of CO_2 : 5 moles of $C_5H_{12} \times \dfrac{5 \text{ moles } CO_2}{1 \text{ moles } C_5H_{12}} = 25$ moles CO_2

(5)   moles of H_2O : 3.0 moles of $C_5H_{12} \times \dfrac{6 \text{ moles } H_2O}{1 \text{ moles } C_5H_{12}} = 18$ moles H_2O

g of H_2O : 18 moles $H_2O \times \dfrac{18.0 \text{ g}}{\text{mol}} = 324$ g $H_2O = 320$ g H_2O

g of CO_2 : 3.0 moles of $C_5H_{12} \times \dfrac{5 \text{ moles } CO_2}{1 \text{ moles } C_5H_{12}} \times \dfrac{44.0 \text{ g } CO_2}{\text{mol } CO_2} = 660$ g

(6)   % yield $= \dfrac{550 \text{ g}}{660 \text{ g}} \times 100 = 83.33 = 83\%$ yield of CO_2

3. (1) (b)   (2) (d)   (3) (c)

4. 100. g of ZnS and 100. g of O_2 used so we first need to convert to moles of each and then compare the stoichiometry.

Moles of ZnS = 100. g of ZnS $\times \dfrac{\text{mole}}{97.4 \text{ g}} = 1.03$ moles ZnS

Moles of O_2 = 100. g of $O_2 \times \dfrac{\text{mole}}{32.0 \text{ g}} = 3.13$ moles O_2

We need 2 moles of ZnS for every 3 moles of O_2; for 1.03 mols of ZnS, at least 1.03 moles ZnS $\times \dfrac{3 \text{ moles } O_2}{2 \text{ moles ZnS}} = 1.55$ moles O_2 would be needed.

Since 1.55 moles of O_2 would be needed and there is 3.13 moles of O_2 available, ZnS is the limiting reagent.

Theoretical moles of ZnO = 1.03 moles ZnS $\times \dfrac{2 \text{ moles ZnO}}{2 \text{ moles ZnS}} = 1.03$ moles ZnO

5. (1)   6.69 g of O_2
The equation must be balanced before we can determine a proper mole ratio for calculating the amount of O_2.

$$2\,PbO_2 \xrightarrow{\Delta} 2\,PbO + O_2(g)$$

Next, we need to convert 100. grams of PbO_2 into the number of moles of PbO_2.

The molar mass of $PbO_2 = 207.2 \text{ g/mol} + (2 \times 16.00 \text{ g/mol}) = 239.2 \frac{g}{mol}$

$$\text{Number of moles of } PbO_2 = (100.\,\cancel{g})\left(\frac{1 \text{ mol}}{239.2 \,\cancel{g}}\right) = 0.418 \text{ mol}$$

$$\text{The mole ratio is } \frac{\text{mol desired substance}}{\text{mol starting substance}} \Longrightarrow \frac{1 \text{ mol } O_2}{1 \text{ mol } PbO_2}$$

$$\text{Moles of } O_2 = (0.418 \,\cancel{\text{mol } PbO_2})\left(\frac{1 \text{ mol } O_2}{2 \,\cancel{\text{mol } PbO_2}}\right) = 0.209 \text{ mol } O_2$$

To convert moles of O_2 into grams, we multiply the number of moles by the molar mass.

$$0.209 \text{ mol} \times 32.00 \frac{g}{mol} = 6.69 \text{ g } O_2$$

(2) 75%
  The percent yield is determined by dividing the actual yield by the theoretical yield, multiplied by 100.

$$\frac{\text{actual}}{\text{theoretical}} \times 100$$

$$\frac{5.0 \,\cancel{g}}{6.7 \,\cancel{g}} \times 100 = 75\%$$

6. (1) $Al(OH)_3$, aluminum hydroxide   (2) 6 moles   (3) 6 moles
    (4) 2 moles $Al_2(SO_4)_3$, 12 moles H_2O

7. The balanced equation is

$$2\,Al + 3\,CuSO_4 \rightarrow 3\,Cu + Al_2(SO_4)_3$$

First we need to determine the number of moles of Cu that can be formed from each reactant.

$$(5.0 \,\cancel{\text{mol Al}})\left(\frac{3 \text{ mol Cu}}{2 \,\cancel{\text{mol Al}}}\right) = 7.5 \text{ mol Cu}$$

$$(10.0 \,\cancel{\text{mol } CuSO_4})\left(\frac{3 \text{ mol Cu}}{3 \,\cancel{\text{mol } CuSO_4}}\right) = 10.0 \text{ mol Cu}$$

Therefore, Al is the limiting reactant and 7.5 mol of Cu can be formed. Next we need to calculate the number of moles of $CuSO_4$ that will react with 5.0 moles of Al.

$$(5.0 \,\cancel{\text{mol } Al})\left(\frac{3 \text{ mol } CuSO_4}{3 \,\cancel{\text{mol } Al}}\right) = 7.5 \text{ mol } CuSO_4$$

Therefore, $10.0 \text{ mol } CuSO_4 - 7.5 \text{ mol } CuSO_4 = 2.5 \text{ mol of } CuSO_4$ in excess.

8. **Solution map:**   Calculate the moles of ZnO produced by each reactant.

$$\text{g ZnS} \rightarrow \text{mol ZnS} \rightarrow \text{mol ZnO}$$

$$\text{g } O_2 \rightarrow \text{mol } O_2 \rightarrow \text{mol ZnO}$$

Whichever reactant produces less mol of ZnO is the limiting reactant and that calculation is the relevant one.

$$(100 \text{ g ZnS})\left(\frac{1 \text{ mol ZnS}}{97.46 \text{ g ZnS}}\right)\left(\frac{2 \text{ mol ZnO}}{2 \text{ mol ZnS}}\right) = 1.03 \text{ mol ZnO}$$

$$(100 \text{ g O}_2)\left(\frac{1 \text{ mol O}_2}{32.00 \text{ g O}_2}\right)\left(\frac{2 \text{ mol ZnO}}{2 \text{ mol O}_2}\right) = 2.08 \text{ mol ZnO}$$

1.03 mole ZnO, ZnS is limiting reactant, O_2 is in excess.

9.   4300 g Fe_2O_3 (2 significant figures)

Solution map: Balance equation

$$\text{kg Fe} \rightarrow \text{g Fe} \rightarrow \text{mol Fe} \rightarrow \text{mol Fe}_2O_3 \rightarrow \text{g Fe}_2O_3$$

The balanced equation is

$$4 \text{ Fe} + 3 \text{ O}_2 \rightarrow 2 \text{ Fe}_2O_3$$

$$\text{The number of moles of Fe} = (3.0 \text{ kg})\left(1000 \frac{\text{g}}{\text{kg}}\right)\left(\frac{1 \text{ mol}}{55.85 \text{ g}}\right) = 54 \text{ mol of Fe}$$

$$\text{moles of Fe}_2O_3 = (54 \text{ mol Fe})\left(\frac{2 \text{ mol Fe}_2O_3}{4 \text{ mol Fe}}\right) = 27 \text{ mol}$$

To determine the number of grams, we multiply 27 moles of Fe_2O_3 by the molar mass.

$$\text{The molar mass of Fe}_2O_3 = (2 \times 55.85 \text{ g/mol}) + (3 \times 16.00 \text{ g/mol}) = 159.7\frac{\text{g}}{\text{mol}}$$

$$27 \text{ mol} \times 159.7\frac{\text{g}}{\text{mol}} = 4300 \text{ g Fe}_2O_3 (2 \text{ significant figures})$$

10.   (1) MX_2

(2)   The mole ratio $X_2:MX_2$ is 1:2 so 0.342 moles of X_2 requires a minimum 0.684 moles of MX_3.

(3)   8 MX_2 theoretically yields 8 MX_3 but only 6 MX_3 were obtained at the end of the reaction. The actual yield is therefore 75%.

11.   2.31 g

2.000 gallons is equal to 7568 mL which has a density of 1.018 g/mL, The mass of the solution therefore is:

$$(1.018 \text{ g/mL})(7568 \text{ mL}) = 7704 \text{ g}$$

Of this mass, 7704 g, 300.ppm are silver ion. To calculate how many grams of silver might be reclaimed with no loss, we set up the following equation.

$$\left(\frac{1}{1 \times 10^6 \text{ ppm}}\right)(300. \text{ ppm})(7704 \text{ g}) = 2.31 \text{ g}(3 \text{ significant figure})$$

12. 0.161 g filter paper

The equation tells us that 1 mole of C reduces 4 moles of Ag ion to metallic silver.

Therefore, moles of carbon required is equal to

$$\left(\frac{2.31 \text{ g Ag}}{107.9 \text{ g/mol}}\right)\left(\frac{1 \text{ mol C}}{4 \text{ mol Ag}}\right) = 0.00535 \text{ mol carbon}$$

We need a piece of filter paper that will contain at least 0.00535 moles carbon. The percentage of carbon in the cellulose is

$$\% \text{ C} = \frac{72.06 \text{ g}}{180.2 \text{ g}} \times 100 = 39.99\%$$

To calculate the mass of filter paper needed we can convert the number of moles of carbon into grams and divide this value by the amount of carbon in the paper.

$$\text{paper mass} = (0.00535 \text{ mol C})(12.01 \text{ g/mol})\left(\frac{1 \text{ g filter paper}}{0.3999 \text{ C}}\right) = 0.161 \text{ g}$$

CHAPTER TEN

Modern Atomic Theory and the Periodic Table

SELECTED CONCEPTS REVISITED

A brief summary of the contributions by some scientists covered in this chapter is:

Max Planck stated that energy is not released continuously, but in quanta.
Niels Bohr proposed that electrons are found in quantized energy levels or orbits.
Louis de Broglie suggested that all objects have wave properties.
Erwin Schroedinger's mathematical model of an electron's wave properties led to the proposal of electrons being found somewhere within a given area called an orbital, and not along a specific path or orbit.

Electron configurations designate in which orbital each electron of the atom is found. Use the periodic table to help you remember the order of filling orbitals. Divide the periodic table into the s-block, d-block, p-block, and f-block as shown later in this section. Notice the blocks follow the natural divisions of the periodic table as written (this is no accident!).

Here is a summary of some things to remember:

Similar orbitals are half-filled before electrons are paired up in those orbitals. For example, the three 2p orbitals in N each "house" one electron. For an O atom, two of the 2p orbitals have a single electron and one has a pair of electrons as opposed to one being empty and the other two having two electrons each.

The d orbitals are filled after the next higher s orbitals. For example, the 3d orbitals are filled after the 4s orbitals. The 5d orbitals are filled after the 6s orbitals.

The f orbitals are two less than the s orbitals. That is, the 4f orbitals are filled after the 6s orbitals.

An example of an orbital diagram $\boxed{\uparrow\downarrow}$ $\boxed{\uparrow\downarrow}$ $\boxed{\uparrow\downarrow}$ $\boxed{\uparrow\downarrow}$ $\boxed{\uparrow}$

An example of an electron configuration $1s^2 2s^2 2p^5$

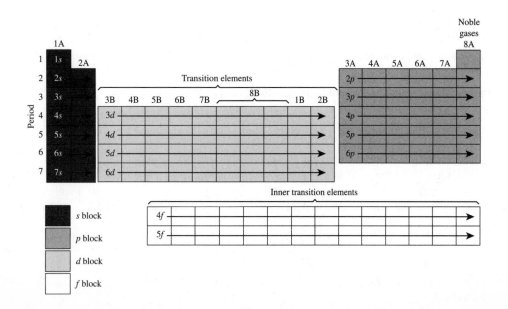

You can easily reduce electron configuration errors by checking to ensure that the sum of the superscripts is equal to the total number of electrons in the atom. For example, F has an electron configuration of $1s^2 2s^2 2p^5$. Sum of the superscripts $= (2 + 2 + 5) = 9$ and an atom of F has 9 electrons so you at least know you assigned the correct number of electrons.

Remember that the 3d level is filled after the 4s level, i.e., $4s^2 3d$ (Not $4s^2 4d$)

$3s^2 3p^6 4s$ NOT $3s^2 3p^6 3d$ – You can avoid making this mistake if you use your periodic table to help you with electron configuration. As you go from left to right across the table until you reach the end of a row and then go to start of the next row, you proceed from the p-block (at the right of the table) to the s-block (at the far left of the table). The d-block comes after the s-block.

RECAP SECTION

In Chapter 10 you learn more about the structure of the atom and the arrangements of the sub-atomic particles. The Bohr model was able to describe hydrogen but not other atoms, so a newer model was devised where electrons occupy regions of space rather than travel on specific pathways. This results in an atomic model of a small dense nucleus, surrounded by a series of electron clouds. The electrons have properties of both particles and waves, so this chapter also presents the basic characteristics of electromagnetic radiation. The electron energy levels are quantized and give rise to unique line spectra for each atom. The knowledge of the principal energy levels and sublevels, and orbitals of an atom are used to write electron configuration. The position of an atom on the periodic table gives information about the electron configuration and vice versa. The periodic table continues to be a valuable tool, used by scientists daily.

SELF-EVALUATION SECTION

1. According to the Bohr model of the hydrogen atom,

    (1) what happens to the electron when it absorbs energy?
    (2) when an electron loses energy, what happens to that excess energy?

2. Fill in the blanks

    (1) The current theory that best explains electron behavior is called the _____ theory.
    (2) This theory states that it is impossible to know the exact position of an electron at any instant of time; instead the region of space where it is most probable to find an electron is called an _____ _____.

3. Review the basic rules and energy level order regarding the state of electrons in atoms.

    a. In the ground state (lowest energy state) of an atom, electrons tend to occupy orbitals of the lowest possible energy.
    b. Each orbital may contain a maximum of two electrons (with opposite spins).
    c. The energy level order is.

    $$1s, \ 2s, \ 2p, \ 3s, \ 3p, \ 4s, \ 3d$$

    The maximum number of electrons that can be found in any main energy level can be determined by a formula involving the term n where n is the (1) _____. When $n = 1$, the total number of electrons will be (2) _____, which will fill the sublevels (3) $s, \ p, \ d, \ f$. For $n = 2$, the total number of electrons will be (4) _____, which will fill the sublevels (5) $s, \ p, \ d, \ f$. Look at the following symbol and describe its meaning:

    $$4s^2$$

    The number 4 represents (6) _____, the small letter s represents (7) _____, and the superscript 2 represents (8) _____.

Now we can write electron configurations. For example, lithium has atomic number 3. Therefore, the correct way of writing the electron configuration would be Li $1s^2 2s^1$.

Try the configurations for

(9) sulfur	Atomic number 16
(10) neon	Atomic number 10
(11) potassium	Atomic number 19
(12) cobalt	Atomic number 27
(13) magnesium	Atomic number 12

Write your answers here.

Determine the atomic number for the element represented by the following electron configurations.

(14) $1s^2 2s^2 2p^6 3s^2 3p^1$ _____

(15) $1s^2 2s^2 2p^3$ _____

(16) $1s^2 2s^2 2p^6 3s^2 3p^6 4s^2 3d^5$ _____

(17) $1s^2 2s^2 2p^6 3s^2 3p^6 4s^2 3d^2$ _____

(18) $1s^2 2s^2 2p^6 3s^2 3p^5$ _____

4. When n = 1, a total of (1) _____ electrons will fill the (2) _____ sublevel.

When n = 2, a total of (3) _____ electrons will fill the (4) _____ (the outermost) sublevel.

Given the symbol $4s^2$, the 4 represents (5) _____, the s represents

(6) _____, and the 2 represents (7) _____.

5. Atomic structures may also be represented by orbital diagrams. Using this method, diagram the electron structure for the following. For the first three, circle the orbitals containing the valence electrons.

(1) $_6$C
(2) $_{19}$K
(3) $_{10}$Ne
(4) $_{23}$V
(5) $_{33}$As
(6) $_{17}$Cl

Write your answers here.

6. Indicate whether the statement made is true or false, or fill in the appropriate blank.

   (1)   A horizontal group of elements in the periodic table is called a _____.

   (2) T/F   All transition elements are metals.

   (3) T/F   Known elements are arranged in a table according to increasing atomic mass.

   (4)   A vertical group of elements in the periodic table is called a _____.

   (5)   The distinguishing electrons for the transition elements are filling the _____ and _____ shells.

   (6) T/F   Elements in the same row have similar chemical properties because they all have the same number of electrons in their outermost energy level.

7. Using the periodic table in Chapter 10 of the textbook, find the following items of information about each element.

	Symbol	Atomic Number	Atomic Mass	No. of Valence Electrons
(1) Phosphorus	_____	_____	_____	_____
(2) Fluorine	_____	_____	_____	_____
(3) Mercury	_____	_____	_____	_____
(4) Cesium	_____	_____	_____	_____

8. Identify the element represented by the orbital diagram.

   (1) ⇅ ⇅ ⇅⇅⇅ ⇅ ↑ ↑ ↑

   (2) ⇅ ⇅ ⇅⇅⇅ ⇅ ⇅⇅⇅ ⇅ ⇅↑ ↑ ↑

9.

   (1) For each of the following, label the appropriate row or column on the periodic chart.
      (a) alkali metals
      (b) halogens
      (c) transition metals
      (d) noble gases
      (e) alkaline earth metals

   (2) On the periodic chart, draw an arrow on Group 4A showing the direction of <u>increasing</u> metallic characteristics.

   (3) On the periodic chart, indicate the s-block elements, the p-block, d-block and f-block elements.

Challenge Problem

10. After reviewing examples in the text for a few minutes, try your hand at writing the complete electronic configuration ($1s$, $2s$, etc.) for (1) $_{21}Sc$ (2) $_{29}Cu$ (3) $_{40}Zr$.

Write your answers here.

ANSWERS TO QUESTIONS AND SOLUTIONS TO PROBLEMS

1. (1)   In the Bohr model, when an electron absorbs energy, it makes a quantized jump to a higher energy level.

   (2)   When an electron loses energy and makes the quantized jump to a lower energy level, the excess energy is given off as light energy.

2. (1) quantum mechanics (or wave mechanics)   (2) electron cloud

3. (1) principal energy level   (2) 2   (3) s (4) 8   (5) s, p
   (6) principal energy level   (7) Type of sublevel   (8) Number of electrons in that sublevel
   (9) $1s^2 2s^2 2p^6 3s^2 3p^4$ (10) $1s^2 2s^2 2p^6$ (11) $1s^2 2s^2 2p^6 3s^2 3p^5$
   (12) $1s^2 2s^2 2p^6 3s^2 3p^6 4s^2 3d^7$ (13) $1s^2 2s^2 2p^6 3s^2$ (14) 13   (15) 7
   (16) 25   (17) 22   (18) 17

4. (1) 2   (2) s   (3) 6   (4) p   (5) principal energy level
   (6) sublevel (s, p, d, f)   (7) the number of electrons in that sublevel

5. (1)   C

   (2)   K

   (3)   Ne

   (4)   V

   (5)   As

   (6)   Cl

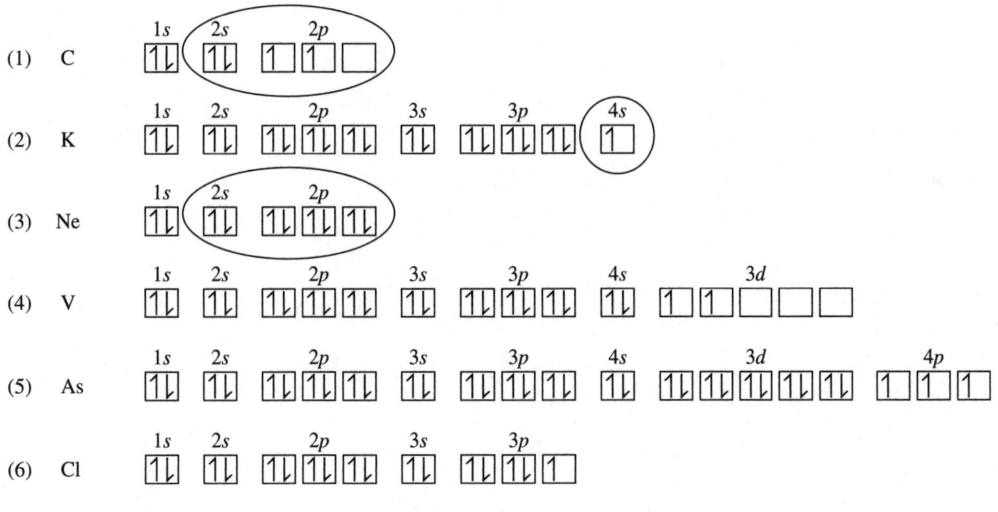

6. (1) row or period   (2) T
   (3) F   (4) column or family
   (5) d, f   (6) F

7. (1) P   15   30.97   5
   (2) F   9   19.00   7
   (3) Hg   80   200.6   2
   (4) Cs   55   132.9   1

8. (1) Phosphorus   (2) Iron

9.

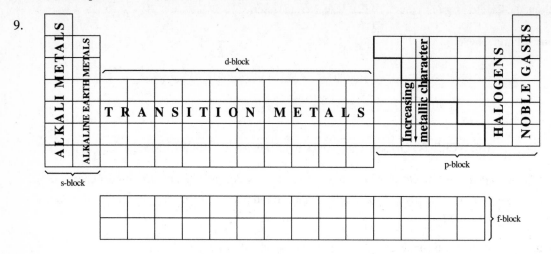

10. (1) $_{21}Sc$ $1s^2 2s^2 2p^6 3s^2 3p^6 4s^2 3d^1$
    (2) $_{29}Cu$ $1s^2 2s^2 2p^6 3s^2 3p^6 4s^1 3d^{10}$
    (3) $_{40}Zr$ $1s^2 2s^2 2p^6 3s^2 3p^6 4s^2 3d^{10} 4p^6 5s^2 4d^2$

SELECTED CONCEPTS REVISITED

$M_{(g)} \rightarrow M^+_{(g)} + e^-$ energy required for this process = IONIZATION ENERGY

Ionization energy (I.E.) – energy required to completely **remove** an electron.

electron more difficult to remove $\Rightarrow$ more energy needed $\Rightarrow$ Higher (larger) I.E.

Comparing: $M_{(g)} \rightarrow M^+_{(g)} + e^-$ to $M^+_{(g)} \rightarrow M^{2+}_{(g)} + e^-$

As each electron is removed, the remaining electrons feel a stronger pull to the nucleus (more positive charge than negative charge now), so *more difficult to remove each subsequent electron*. It is even more difficult to remove a core electron than a valence electron.

Electrons closer to nucleus experience stronger attractive forces $\Rightarrow$ harder to remove those electrons

   In general: smaller atomic radius $\Rightarrow$ higher ionization energy

Lewis structures show valence electrons. The approximate distribution of the electrons in the valence orbitals is shown, that is, whether they are paired or not.

Ionic bonds are the result of electrostatic interaction (that is, opposite charges attract). Ions (charged species) are formed when electrons are transferred. Ionic bonds involve cations and anions.

Parent atom $\Rightarrow$ cation; atom LOSES e^-, so $\#p > \#e^-$; cationic radius SMALLER than parent atomic radius

Parent atom $\Rightarrow$ anion: atom GAINS e^-, so $\#e^- > \#p$; ; anionic radius LARGER than parent atomic radius

Covalent bonds are the result of a pair of electrons being shared between two atoms. When one atom involved in a covalent bond has a stronger attraction for the shared pair of electrons than the other atom, the result is a polar covalent bond. That is, the electrons are being shared but not equally and because they are more likely to be found nearer the more electronegative atom, that atom will have a partial negative charge. The less electronegative atom will have a partial positive charge.

Notice that the most electronegative elements are the nonmetals. One rationale for the trend in electronegativity is that the smaller the atom, the closer to the nucleus the shared electrons will be and so the more the nucleus is able to attract the electrons. Fluorine is the most electronegative element.

A molecule cannot be polar unless it contains polar bonds. However, a molecule that contains polar bonds is not necessarily polar. To determine if a molecule is polar, you need to look at the shape of the molecule. If the polar bonds are able to cancel each other out, the molecule is not polar. (You can imagine these polar bonds as miniature tug-of-wars. If there can be a winner, the molecule is polar.)

VSEPR theory allows you to predict the shape of a molecule. The idea is that lone pairs and bonds all contain electrons and like charges repel so these areas of negative charges repel each other such that the bonds and lone pairs try to get as far apart from each other as possible. (If you use a balloon to represent a pair of electrons, whether a bonding pair or a nonbonding pair, and tie four balloons together, you will see the balloons point to the four corners of a tetrahedron.)

The number (and therefore arrangement) of electron pairs determine the **geometric shape** of the molecule; the position of the atoms around the center atom determines the **molecular shape** of the molecule. The latter is generally what most people think of when asked for the shape of the molecule. Be sure to clarify what you are being asked for! Geometric shape and molecular shape are sometimes referred to as electronic geometry and molecular geometry respectively.

COMMON PITFALLS TO AVOID

Electrons are negatively charged particles, NOT negatively charged atoms. If atoms <u>gain</u> electrons, they become negatively charged (anions). If atoms <u>lose</u> electrons, they become positively charged (cations).

Do not confuse valence electrons with all electrons in an atom. Electrons in the inner shells are generally not involved in chemical reactions. Only the outermost, i.e. valence, electrons are of concern in Lewis structures.

In general, atomic radius **DECREASES** from LEFT TO RIGHT across the periodic table.

Electronegativity and ionization energy are NOT the same. Ionization energy involves <u>removal of an electron;</u> electronegativity is indicative of how well a <u>shared electron is attracted</u> to an element's nucleus. *Memorizing the definitions (rather than the trend) will help you figure out and understand trends*.

When drawing Lewis structures of covalent species, it is helpful to assign the nonbonding pairs to the outer atoms first. Then if the electrons "run out" before the central atom has its octet, simply convert a nonbonding pair on an outer atom into a bonding pair between that atom and the central atom. Double bonds and extra electrons are generally around the central atom. Also, remember it is often easy to find a mistake in the Lewis structure of a covalent species. If each atom does not have eight electrons (other than H) and the total number of electrons is not equal to the total number of valence electrons of those atoms, then you have likely made a mistake.

Do not forget to use VSEPR to determine the shape of the molecule **before** trying to decide if the molecule is polar.

RECAP SECTION

Knowing the definitions of terms, and the understanding the organization of the periodic table, will enable you to derive many of the periodic table trends so that you don't have to memorize them all. Understanding what gives rise to the atomic trends and knowing definitions such as ionization energy will allow you to figure out the trends without memorizing them. Knowing how to read the periodic table and the information inherent in it will make it easier to draw Lewis structures, and predict which bonds are ionic, polar covalent, or covalent. Drawing the correct Lewis structure of a compound gives information about its shape and polarity, and whether resonance structures exist, which will eventually lead to a better understanding of their chemical reactivity.

SELF-EVALUATION SECTION

1. Fill in the blank space or circle the appropriate response.

    When a neutral atom loses an orbital electron, it becomes a (1) <u>positively/negatively</u> charged ion. The energy required to remove a mole of electrons from a mole of atoms is called the (2) _____ energy and is a relatively (3) <u>low/high</u> value for Group I and Group II metals and (4) <u>low/high</u> for non-metallic elements. The ionization energy for the noble gas elements is especially (5) <u>high/low,</u> indicating that eight electrons in the valence shell of an atom is a very stable structure. If you list the ionization energies of the elements from a group in the periodic table, the element from the top of the group has a (6) <u>higher/lower</u> ionization energy than the element at the bottom of the group. Two factors account for this experimental observation. As you go down a group, the electron being removed is (7) <u>farther from/closer to</u> the nucleus, and the increasing number of filled electron orbitals shields the valence electrons from the positive nucleus. The valence electrons are shown in the Lewis structures. We will use Lewis structures in the next section to illustrate the concept of forming chemical bonds.

    As mentioned above, when a neutral atom (8) <u>loses/gains</u> an electron it becomes positively charged. A positively charged ion is also called a (9) _____. For example, a potassium atom has 19 protons and 19 electrons.

A potassium ion has a charge of +1, which means the ion has (10) _____electrons instead of 19, as in the neutral atom. Conversely, a negatively charged ion, which is called an (11) _____, is formed when a neutral atom (12) loses/gains electrons. A chlorine atom has 17 protons and 17 electrons and becomes an anion by (13) gaining/losing an electron. A chloride ion has a stable outer shell of (14) _____electrons. A sodium atom in close proximity to a chlorine atom can also reach a stable outer shell of eight electrons by losing one electron. Thus, both sodium and chlorine reach a stable electron structure by the process of electron transfer. The metallic elements attain a stable structure by (16) gaining/losing electrons; the nonmetallic elements attain a stable structure by (17) gaining/losing electrons.

2. Using Lewis structures, illustrate how the following compounds are formed. Some are electron-transfer problems and some are electron-sharing problems. You may need to consult the periodic table found in the front of the book.

   (1)   zinc iodide (ZnI_2)

   (2)   potassium oxide (K_2O)

   (3)   water (H_2O)

   (4)   silicon dioxide (SO_2)

   (5)   sulfur trioxide (SO_3)

   (6)   phosphorus trichloride (PCl_3)

3. Write out the Lewis structure for HNO_2, nitrous acid. You may use a periodic table and follow problems in the text as a guide.

4. Draw Lewis structures for the following:
   (1) PO_4^{3-} (2) NH_4Br (3) ClO^-
   (4) MnO_4^- (5) H_2S (6) CO_3^{2-}

5. If we examine many different compounds for the kinds of chemical bonds that hold them together, we will generally find two types. The two types of chemical bonds are called (1) _____ and the (2) _____bond. When a transfer of electrons takes place from one atom to another, the resulting charged particles form an (3) _____ bond.

   A cation and an anion will form an (4) _____ bond since oppositely charged particles (5) attract/repel each other. Metallic elements tend to form ionic bonds when combining with the nonmetals, as we saw in the previous section.

   The predominant type of chemical bond is the (6) _____bond. This type of bond occurs in the hydrogen molecule and develops as a result of each hydrogen atom contributing (7) one/two electron(s) to form the bond. The $1s$ electron orbitals of the hydrogen atoms overlap and pair to form a stable hydrogen molecule. There is a strong tendency for the hydrogen molecule to form from two individual atoms since in the molecule each (8) _____ charged electron is attracted to two (9) _____ charged nuclei.

   The covalent bond is usually indicated by a dash mark (—). A single dash means (10) _____ pair of electrons and a double dash means (11) _____ pairs of electrons.

6. Given the following table of electronegativity values, indicate which of the listed binary compounds has polar covalent bonds. Also, calculate the difference in electronegativity values for one of the covalent bonds in each molecule.

Electronegativity Values

H 2.1	Br 2.8	Se 2.4
B 2.0	Cl 3.0	Te 2.1
P 2.1	F 4.0	N 3.0
O 3.5	S 2.5	Na 0.9

For example, the compound HF has one covalent bond. Is the bond between H and F polar? Yes, there is a significant difference in electronegativity values of $4.0 - 2.1 = 1.9$. Examine the remaining compounds in the same manner.

		Polar Covalent Bond **(yes or no)**	**Electronegativity** **Value Difference**
(1)	NCl_3	_____	_____
(2)	BrCl	_____	_____
(3)	IBr	_____	_____
(4)	H_2Se	_____	_____
(5)	H_2O	_____	_____
(6)	Br_2	_____	_____
(7)	OF_2	_____	_____
(8)	PH_3	_____	_____
(9)	H_2Te	_____	_____
(10)	NH_3	_____	_____
(11)	H_2S	_____	_____

7. Based on the electronegativity value difference, which of the following pairs would you expect to have a more polar covalent bond?

   (1) BrCl or OF_2

   (2) H_2Se or H_2O

8. In the following, which of each pair will be larger?

   (1) Br^- or Br   (3) Na^+ or Na
   (2) Cu^{2+} or Cu^+ (4) Fe^{2+} or Fe^{3+}

9. (1) Draw the Lewis structure for CO_2.

   (2) Are the bonds in CO_2 polar covalent?

   (3) Is the molecule CO_2 polar or non-polar?

10. Predict the type of bond that would be formed between the following pairs of atoms. Use Table 11.5 in the text.
    (1) Li & I   (2) C & H   (3) Al & N
    (4) Cs & P   (5) Se & S   (6) Ba & C

ANSWERS TO QUESTIONS AND SOLUTIONS TO PROBLEMS

1. (1) positively   (2) ionization   (3) low   (4) high
   (5) high   (6) higher   (7) farther from   (8) loses
   (9) cation   (10) 18   (11) anion   (12) gains
   (13) gaining   (14) eight   (15) losing   (16) gaining

2. (1) Zinc is a Group 2B element, so it has two valence electrons in its outer shell. Iodine is a Group 7A element, so it has seven valence electrons in its outer shell. Zinc loses one electron to each of the iodine atoms and becomes a +2 charged cation. Each iodide ion thus has a −1 charge.

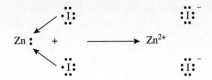

(2)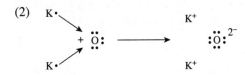

(3)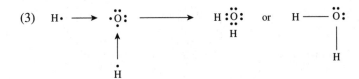

(4) In order to write a Lewis structure for the molecule SiO₂, which will have eight electrons around each atom, we must use double bonds between Si and each oxygen atom. Each double bond consists of four electrons – two from the Si atom and two from each oxygen atom.

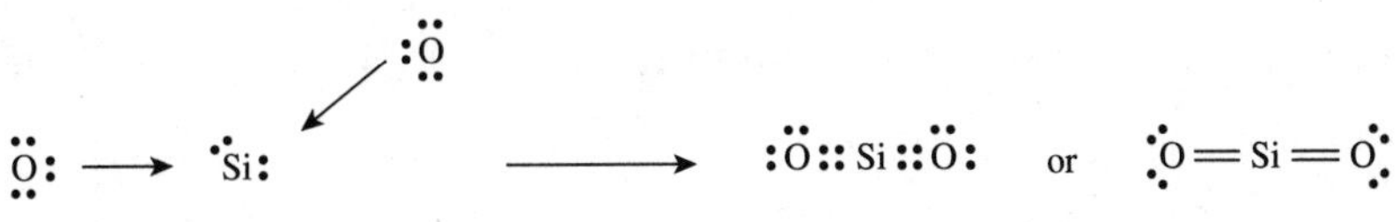

The problem involves the formation of double bonds. Since one pair of electrons is one bond, there are two bonds between the two oxygen atoms and the silicon atom. Only in this manner will each atom have eight electrons in its outer shell.

(5) 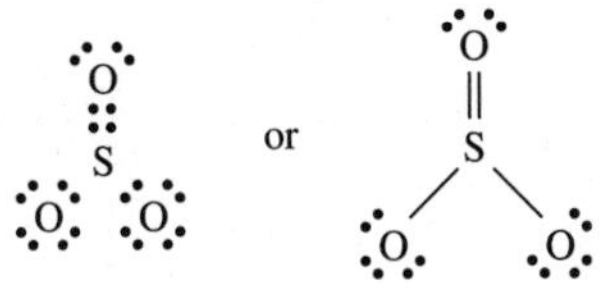 *or*

(6) The Lewis structure for SO₃ involves one double bond and two single bonds in order to place eight electrons around each atom.

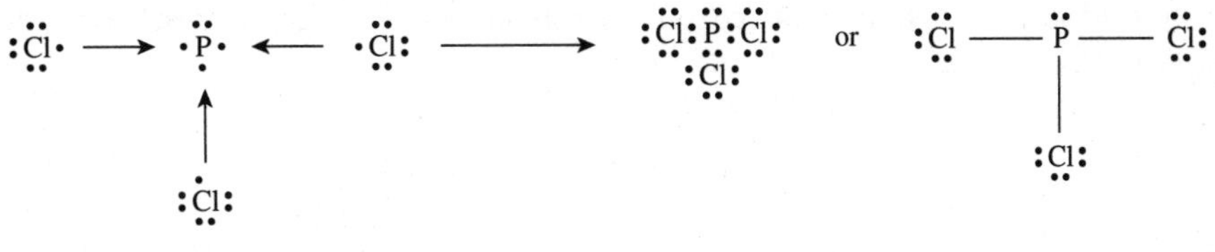

3. The first step was to count the total number of valence electrons in the molecule.
valence electrons $= 1 + 5 + 6 + 6 = 18$

Next write the skeletal structure of the atoms and connect them with a single bond (two electrons). How do we know the arrangement of the atoms? One clue is that HNO₂ is an acid, which suggests the H is bonded to an O.

H:O:N:O or H—O—N—O

Then figure out how many electrons are left to be assigned.

Started with 18 electrons and used two for each bond shown. $18 - 6 = 12$. 12 electrons left

Now distribute pairs of electrons around each atom (nonbonding pairs) resulting in noble gas configurations. Note that H only needs a total of 2 electrons to achieve "noble gas status". Most of the rest of the elements need 8 electrons. (Hint, assign nonbonding pairs to "outer" elements first.)

$$H \cdot\cdot \ddot{\ddot{O}} \cdot\cdot \ddot{N} \cdot\cdot \ddot{\ddot{O}}: \qquad \text{or} \qquad H - \ddot{\ddot{O}} - \ddot{N} - \ddot{\ddot{O}}:$$

The twelve remaining electrons are used (i.e., all 18 have now been assigned), but not all the atoms are satisfied. The H has two electrons (good), the oxygens each have 8 electrons (good), but the nitrogen only has 6 electrons around it; it wants two more.

So the penultimate step is to shift a nonbonding pair of electrons from an outer atom to make a double bond. In other words, share a pair of electrons with the central atom.

$$H \cdot\cdot \ddot{\ddot{O}} \cdot\cdot \ddot{N} \cdot \ddot{\ddot{O}}: \qquad \text{or} \qquad H - \ddot{\ddot{O}} - \ddot{N} - \ddot{\ddot{O}}:$$

And the last step is to always double-check! 18 electrons total with 8 electrons around each atom (except for H).

$$H \cdot\cdot \ddot{\ddot{O}} \cdot\cdot \ddot{N} :: \ddot{O}: \qquad \text{or} \qquad H - \ddot{\ddot{O}} - \ddot{N} = \ddot{O}:$$

4. (1) $\left[\begin{array}{c} :\ddot{O}: \\ :\ddot{O}:\overset{..}{P}:\ddot{O}: \\ :\ddot{O}: \end{array} \right]^{3-}$ or $\left[\begin{array}{c} :\ddot{O}: \\ | \\ :\ddot{O}-P-\ddot{O}: \\ | \\ :\ddot{O}: \end{array} \right]^{3-}$

(2) $\left[\begin{array}{c} H \\ \overset{..}{} \\ H:N:H \\ H \end{array} \right]^{+}$ $\left[:\ddot{\underset{..}{Br}}: \right]^{-}$ or $\left[\begin{array}{c} H \\ | \\ H-N-H \\ | \\ H \end{array} \right]^{+}$

(3) $\left[:\overset{..}{\underset{..}{Cl}}:\ddot{O}: \right]^{-}$ or $\left[:\overset{..}{\underset{..}{Cl}} - \ddot{O}: \right]^{-}$

(4) $\left[\begin{array}{c} :\ddot{O}: \\ :\ddot{O}:Mn:\ddot{O}: \\ :\ddot{O}: \end{array} \right]^{-}$ or $\left[\begin{array}{c} :\ddot{O}: \\ | \\ :\ddot{O}-Mn-\ddot{O}: \\ | \\ :\ddot{O}: \end{array} \right]^{-}$

(5) $H:\ddot{\underset{..}{S}}:H$ or $H-\ddot{\underset{..}{S}}-H$

(6)

$\left[\begin{array}{c} \cdot\ddot{O}\cdot \\ :: \\ C \\ :\ddot{O}: \ :\ddot{O}: \end{array} \right]^{2-}$ or $\left[\begin{array}{c} \cdot\ddot{O} \\ || \\ C \\ / \ \backslash \\ \cdot\ddot{O} \quad \ddot{O}\cdot \end{array} \right]^{2-}$

There are 4e⁻ from the carbon, 6e⁻ from each oxygen and 2 additional (negative charge on the ion is -2). So we have to distribute 24e⁻ between the central carbon and the 3 oxygen atoms. This means that one of the bonds needs to be a double bond.

5. (1) ionic   (2) covalent   (3) ionic
    (4) ionic   (5) attract   (6) covalent
    (7) one   (8) negatively   (9) positively
    (10) one   (11) two

6.

Compound	Polar Covalent Bond (yes or no)	Electronegativity Value Difference
(1) NCl₃	no	0
(2) BrCl	yes	0.2
(3) IBr	yes	0.3
(4) H₂Se	yes	0.3
(5) H₂O	yes	1.4

(6)	Br_2	no	0
(7)	OF_2	yes	0.5
(8)	PH_3	no	0
(9)	H_2Te	no	0
(10)	NH_3	yes	0.9
(11)	H_2S	yes	0.4

7.   (1) OF_2 (BrCl difference is 0.2, OF_2 difference is 0.5)

      (2) H_2O (H_2Se difference is 0.3, H_2O difference is 1.4)

8.   (1) Br^- will be larger   (3) Na will be larger

     (2) Cu^+ will be larger   (4) Fe^{2+} will be larger

9.   (1) CO_2 has $(4+6+6) = 16$ valence electrons. Its Lewis structure is

$$\overset{..}{\underset{..}{O}} = C = \overset{..}{\underset{..}{O}}$$

      (2)   Both sets of double bonds between C and O are polar because of the difference in the electronegativities of C and O.

      (3) CO_2 is a non-polar molecule. Although each of the double bonds is polar, the molecule is linear, so the polarities negate each other.

10.   (1) polar covalent   (2) polar covalent   (3) polar covalent

     (4) polar covalent   (5) nonpolar covalent   (6) ionic

WORD SEARCH 1

In the given matrix of letters, find the terms that match the following definitions. The terms may be horizontal, vertical, or on the diagonal. They may also be written forward or backward. Answers are found at the end of Chapter 20.

1. A subatomic particle with a charge of $+1$.
2. The mass of an object divided by its volume.
3. The basic building block of matter that cannot be broken down into simpler substances by ordinary chemical changes.
4. An electrically charged atom or group of atoms.
5. The central part of an atom.
6. An element that is ductile and malleable.
7. The abbreviation for the name of an element.
8. The chemical law concerning the occurrence of the chemical properties of elements.
9. State of matter that is least compact.
10. Atoms of an element having the same atomic number but different atomic masses.
11. A subatomic particle with a charge of -1.
12. A negatively charged particle.
13. A small, uncharged individual unit of a compound.
14. Matter having uniform properties throughout.
15. The smallest particle of an element that can enter into a chemical reaction.
16. The relative attraction that an atom has for the electrons in a covalent bond.
17. A molecule with a separation of charge.
18. A subatomic particle that is electrically neutral.
19. The energy required to remove an electron from an atom.
20. A substance composed of two or more elements combined in a definite proportion by mass.
21. A solid without definite crystalline form.
22. A cloud-like region around the nucleus where electrons are located.

23. The metallic elements characterized by increasing numbers of *d* and *f* electrons in an inner shell.
24. Metric unit of length.

H	C	O	M	P	O	U	N	D	D	I	S	O	T	O	P	E	S
O	X	C	S	Z	K	L	O	Y	P	M	N	Q	R	B	L	C	T
M	P	R	U	D	K	Q	T	V	F	V	G	M	W	E	Z	T	Y
O	I	Z	E	O	R	V	O	M	T	S	E	O	C	U	A	L	G
G	U	A	L	T	E	I	R	O	R	E	S	T	T	L	A	C	I
E	I	F	C	S	U	P	P	W	L	L	R	A	B	A	I	T	U
N	G	O	U	Q	E	G	Y	O	H	O	J	Y	G	R	N	O	P
E	E	E	N	E	U	T	R	O	N	P	A	M	V	E	E	S	E
O	L	B	U	I	S	W	H	E	J	I	M	N	M	T	K	O	R
U	S	O	V	T	Z	U	G	J	F	D	V	E	C	L	D	E	I
S	I	E	I	R	K	A	O	A	B	K	L	T	U	M	L	E	O
K	N	G	S	D	T	M	T	H	V	E	P	R	S	D	D	A	D
L	L	A	T	I	B	R	O	I	P	A	D	O	I	M	E	R	I
O	E	W	V	S	L	H	F	L	O	R	G	E	U	Q	N	V	C
B	O	I	N	N	A	P	A	E	N	O	L	T	W	S	Z	L
M	T	V	M	E	O	S	I	D	N	C	E	M	N	O	I	N	A
Y	L	S	A	T	R	T	Z	M	A	B	U	N	A	L	T	U	W
S	P	J	D	A	T	C	I	O	T	M	J	L	E	K	Y	V	Z
A	E	O	R	B	C	M	R	P	E	B	U	N	A	L	T	U	W
B	H	T	L	C	E	A	H	T	M	X	Y	U	T	T	G	G	C
G	D	I	U	F	L	Q	A	O	F	E	S	N	X	G	A	Y	V
S	T	N	E	M	E	L	E	N	O	I	T	I	S	N	A	R	T
A	M	G	W	U	P	Q	N	S	W	D	L	E	R	V	K	C	B
H	I	V	R	E	T	E	M	O	R	D	Y	H	R	W	E	J	F

CHAPTER TWELVE

The Gaseous State of Matter

SELECTED CONCEPTS REVISITED

Although this chapter does involve important calculations relating to gases, you should first master understanding qualitatively how gas molecules move, how they exert pressure on a container, and how they are affected by temperature. Once you picture how molecules affect and are affected by their surroundings, you will be much better able to decide which calculation to use and use it properly. Pressure, whether it's water pressure or gas pressure, is the force exerted on a unit area by a substance. As gas molecules collide with the container walls, they exert a certain pressure. What will happen to the pressure of a gas sample if more gas molecules are placed in a container? More molecules mean more collisions with the walls. Therefore, the pressure will rise. If the temperature and volume of the container are kept constant, there is a direct relationship between pressure and number of gas molecules. Thus, doubling the amount of gas will double the pressure. Reducing the amount of gas to one-third the original quantity will reduce the pressure correspondingly.

The kinetic molecular theory says that ideal gases are composed of tiny massless particles with no attraction for each other that move in straight lines and lose no energy when they collide with the walls of the container.

The average kinetic energy of a gas is dependent on its temperature and not on the size of the particle. The mass of the particle affects its velocity not its energy. Smaller particles move faster, but larger particles have more mass hitting the container wall, so the pressure exerted on the walls of a container is dependent on the temperature, that is, the average kinetic energy of the particle, not on the specific identity of the particle itself.

The pressure exerted on a container is due to the collisions of the gas molecules on the container. More collisions lead to higher pressures. The number of collisions is affected by the number of particles present, their speed (once they hit one wall, how long before they reach another wall), and the volume of the container (how far do the particles have to travel before they reach another wall). So the pressure of a gas is dependent on the moles of gas present, the temperature of the gas, and the volume of the container.

What happens to a gas sample if the pressure is not constant and the volume cannot expand? Visualize a closed empty can placed on a fire. The temperature of the gas increases; kinetic energy and pressure increase. The gas cannot expand so the pressure increases to the point at which the mechanical strength of the can cannot withstand the high pressure, and the can explodes.

The kinetic molecular theory assumes that the particles of a gas are independent of each other. In a mixture of gases, we can treat each gas in the mixture as though it were the only gas present. So to find the total pressure exerted on the container, we can simply find the pressure exerted by each gas and add them. That is, the total pressure is the sum of the partial pressures. Alternatively, if you were given the total number of moles of all the gases present, we simply find the total pressure and not really worry about the fact that there are different gases present.

Unlike solids and liquids, the density of a gas changes significantly with changes in temperature or pressure. For a given mass of a gas, its volume is highly dependent on its temperature and pressure and therefore its density is affected. So although we can generalize and say that the density of liquid water is 1 g/ml, if we give the density of steam, we would need to specify at what temperature and pressure the density was measured. The density of gases generally uses units of g/L.

When gases react at constant temperature and pressure, the coefficients of a balanced equation can refer to either the volumes of the gas used or the moles of gas used. This again falls back on the fact that regardless of the identity of a gas, at a specific temperature and pressure, one mole of a gas will occupy a specific volume. At STP, that volume (the molar volume) is 22.4 dm^3 (liters).

For most calculations, using 273 instead of 273.15 to convert between K and °C is acceptable.

Be careful when using the ideal gas constant R in calculations. Make sure that the units of R that you use are consistent with your other values. Also be extremely careful to check the units of temperature used in the calculations! Temperature must be in Kelvin for all calculations involving gases.

Room temperature and STP are not the same. STP is an abbreviation for standard temperature and pressure. Standard temperature is 273 K (or 0°C). Room temperature is generally assumed to be 298 K (or 25°C). Standard pressure is 1 atm.

$PV = nRT$ is not the answer to all of your problems! Whenever you do a calculation also try to see if the answer makes sense qualitatively. However, remembering the ideal gas equation can help you remember certain relationships. From $PV = nRT$ you know that $P \propto 1/V$ (assuming the rest of the variables are constant), so if pressure is increased, the volume must have decreased. You can often use logic to get an idea of the relative quantity of the answer even before you start to plug numbers into your calculator.

RECAP SECTION

Chapter 12 is all about gases. Gases lend themselves to visualization on the molecular level. You can picture them in your mind as tiny balls flying around a container. The variables associated with a gas in a container include how many gas particles are present (moles), how large a container (volume), how fast the particles are moving and with how much energy (temperature), and how often and how hard the particles are hitting the walls of the container (pressure). The units used for these measurements may vary and you should be comfortable converting between them as necessary. Boyle's Law relates pressure and volume of a gas at a constant temperature, while Charles' Law looks at the relationship between temperature and volume of a gas at constant pressure. Know the combined, or Ideal Gas Law of $PV = nRT$. Dalton's Law of Partial Pressure works on the premise that the identity of a gas is not important, so the pressure exerted by a specific gas in a mixture is directly proportional to its mole fraction. Use of the ideal gas law solves problems involving pressure, volume, temperature and number of moles, and material from earlier chapters can be incorporated to find the density of a particular gas, and to solve stoichiometric problems relating to gases.

SELF-EVALUATION SECTION

1. The science of chemistry on a quantitative basis began with the systematic study of gas behavior, which may seem somewhat ironic since substances in the gas state are difficult to handle and use for experimental purposes. The fundamental properties of gases were established as compressibility, diffusion, pressure, and expansion. In addition, equal volumes of gases at the same temperature were found to exert identical pressures.

   The assumptions of the kinetic-molecular theory for an ideal gas are listed below.

   (a) Gases consist of tiny particles.

   (b) The volume of gas is mostly empty space.

   (c) Gas molecules have no attraction for each other.

   (d) Gas molecules move in straight lines in all directions, undergoing frequent collisions.

   (e) No energy is lost through the collisions of gas molecules.

   (f) The average kinetic energy for molecules is the same for all gases at the same temperature.

   Properties – list the letter for one or more assumptions from the above list that describe each of the following properties of gases.

   (1) Compressibility   _____

   (2) Pressure   _____

   (3) Expansion   _____

   (4) Pressure of equal volumes of gases   _____

2. Boyle's Law and Charles' Law.
   Given below are several ways of mathematically stating Boyle's law and Charles' law. Place a "B" (for Boyle's) or "C" (for Charles') by each formula. The symbol "$\propto$" means "to vary" or "to be proportional to".

   (1) $PV = k$

   (2) $\dfrac{V_1}{T_1} = \dfrac{V_2}{T_2}$

   (3) $V \propto \dfrac{1}{P}$

   (4) $P_1V_1 = P_2V_2$

   (5) $V \propto T$

   (6) $\dfrac{V}{T} = k$

3. Choose the correct measurement for the following commonly used/referred to quantities.

   (1) _____ standard temperature   (2) _____ molar volume   (3) _____ standard pressure

   (a) 24.0 L   (b) 76 mmHg   (c) 22.4 L

   (d) 273 K   (e) 1 atm   (f) 298 K

4. Make the best match between the Law/quantity and their definitions/quantities.

   (1) _____ Dalton's Law

   (2) _____ Boyle's Law

   (3) _____ Ideal Gas Law

   (4) _____ Avogadro's Law

   (5) _____ Charles' Law

   (a) Equal volumes of two gases at the same P and T will contain equal numbers of particles.
   (b) As the volume of a gas decreases at constant T, its pressure increases.
   (c) The volume of a gas is directly proportional to T in Kelvin at constant P.
   (d) The result of the compilation of Boyle's, Charles', Avogadro's and Gay-Lussac's Laws.
   (e) The pressure of a mixture of gases is simply the sum of their partial pressures under the same conditions.

5. **Gas Law Problems** – Before solving each of these problems, first list the known quantities and then decide (plan) which Law or equation can be used to solve the problem.

   (1) A certain gas exerted a pressure of 145 torr in a 2.00 L container. The gas was compressed into a 675 mL gas bottle. What was the new pressure exerted by the gas?

   Knowns:
   Law/Equation:
   Calculations:

(2)   110. ml of a gas are obtained from a reaction at 240.°C. What volume will the gas occupy at 100.°C?

Knowns:
Law/Equation:
Calculations:

(3)   The pressure of a gas in a piston is 1400. torr. The pressure is then reduced to 650. torr. If the original volume occupied by the gas was 1.00 L, what is the new volume?

Knowns:
Law/Equation:
Calculations:

(4)   What is the volume of a gas brought to STP conditions after being in a 50. mL container at 0.75 atm and 21°C?

Knowns:
Law/Equation:
Calculations:

6.   The density of a gas is expressed in grams per liter.

$$d = \frac{mass}{volume} = \frac{g}{L}$$

At STP, density may also be calculated from the following equation.

$$\text{density at STP} = \frac{\text{molar mass}}{22.4\,\text{L/mol}} = \frac{\text{g/mol}}{\text{liters/mol}} = \frac{g}{L}$$

(1)   At STP, what is the density of CO(carbon oxide) gas? Use your table of atomic masses.

Do your calculations here.

(2)   The density of HCN (hydrogen cyanide) gas at STP is 1.21 g/L. What volume will 100. g of HCN occupy at STP?

Do your calculations here.

(3) A quick way to prepare acetylene gas in the laboratory is to drop chunks of calcium carbide into water. Old-fashioned miners' lamps used this reaction, as did some residential lighting systems years ago. We will work a Dalton's partial pressure problem from this reaction.

A sample of C_2H_2 (acetylene) gas collected over water at 21°C and 750 torr pressure occupies a volume of 175 mL. Calculate the volume of dry acetylene at STP. The vapor pressure of water at various temperatures is listed in Appendix VI of the textbook. Dalton's law states that, in a mixture of gases, each gas exerts its own individual pressure.

$$P_{total} = P_A + P_B + P_C$$

For this particular problem, the total pressure of the moist acetylene collected is made up of two parts – the pressure of C_2H_2 and the pressure of the water vapor. We have to subtract the partial pressure of the water vapor from the C_2H_2 before we correct the C_2H_2 volume to STP.

Do your calculations here.

7. Using the ideal gas equation, calculate the volume that 15 g of H_2 gas at 25°C and 1.2 atm pressure will occupy.

$$PV = nRT$$

Do your calculations here.

8. What volume of NO gas at STP can be produced from 7.0 moles of nitrogen dioxide (NO_2) according to the following equation?

$$3\,NO_2 + H_2O \rightarrow 2\,HNO_3 + NO$$

Do your calculations here.

9. What volume of NO gas will be produced from 4.00 moles of N_2 and 3.00 moles of O_2 at STP according to the following equation?

$$N_2(g) + O_2(g) \rightarrow 2\,NO(g)$$

Do your calculations here.

Challenge Problems

10. The size cylinder known as a 1A cylinder has a volume 43.8 L. How many moles of oxygen are contained in a 1A cylinder at 72°F and 1500 pounds per in^2 pressure (psi)? To convert psi to atmospheres multiply psi by 0.06805 atm/psi. What is the mass of this number of moles of oxygen gas?

Do your calculations here.

11. Use the ideal gas equation to calculate the molar mass of arsine, given that 3.48 g of arsine at 740. torr and 21°C occupies 1.11 L. If arsine is composed of a metal combined with three hydrogen atoms, what is its chemical formula?

Do your calculations here.

ANSWERS TO QUESTIONS AND SOLUTIONS TO PROBLEMS

1. (1) Compressibility – b
   (2) Pressure – d
   (3) Expansion – b, c, d
   (4) Pressure of equal volumes of gases – e, f

2. (1) B   (2) C   (3) B   (4) B   (5) C   (6) C

3. (1) d   (2) c   (3) e

   Note that 24.0 L is the molar gas volume at <u>room</u> temperature. Room temperature is generally considered to be 298 K (25°C). Standard pressure of 1 atm is the same as 760 mmHg not 76 mmHg.

4. (1) e   (2) b   (3) d   (4) a   (5) c

5. (1) 430. torr
       First, write down the known quantities:

       $P_1 = 145.$ torr $P_2 =$ solve for this
       $V_1 = 2.00$ L $V_2 = 675$ mL $= 0.675$ L

       The problem involves only pressure and volume changes, so Boyle's Law may be used. Note: volume decreases so pressure increases

   $$P_1V_1 = P_2V_2 \qquad \text{rearranging to solve for } P_2: \qquad P_2 = \frac{P_1V_1}{V_2}$$

   solve: $\qquad V_2 = \dfrac{(145.\text{ torr})(2.00\,L)}{0.675\,L} = 429.6 \text{ torr} = 430. \text{ torr}$

   (2) 80.0 mL

       Knowns:

       $T_1 = 240.°C = 513K$ $T_2 = 100.°C = 373K$
       $V_1 = 110.$ mL $V_2 =$ what we are trying to solve for

       NOTE: For gas law calculations, temperatures must be in Kelvin, so it is useful to immediately convert them so that you don't forget.

The problem involves only temperature and volume changes, so Charles' Law may be used. Note: temperature decreases so volume decreases

$$\frac{V_1}{T_1} = \frac{V_2}{T_2} \qquad \text{rearranging to solve for } V_2: \qquad V_2 = \frac{V_1 T_2}{T_1}$$

solve: $\qquad\qquad\qquad\qquad V_2 = \frac{(110.\,\text{mL})(373\,\cancel{K})}{513\,\cancel{K}} = 79.98\,\text{L} = 80.0\,\text{mL}$

(3)   2.15 L

Knowns:

$P_1 = 1400.$ torr $P_2 = 650.$ torr
$V_1 = 1.00$ L $V_2 =$ solve for this

The problem involves only pressure and volume changes, so Boyle's Law may be used.
Note: pressure decreases so volume increases

$$P_1 V_1 = P_2 V_2 \quad \text{rearranging to solve for } V_2: \quad V_2 = P_1 V_1 / P_2$$

solve: $\qquad\qquad\qquad\qquad V_2 = \frac{(1400.\,\cancel{\text{torr}})(1.00\,\text{L})}{650.\,\cancel{\text{torr}}} = 2.15\,\text{L}$

(4)   41 mL

Knowns: Reminder: STP conditions are 1 atm and 0°C.

$\qquad\qquad\qquad P_1 = 0.75$ atm $P_2 = 1$ atm (exact so sig. figs. are not limited to 1)

$V_1 = 50.$ mL $V_2 =$ solve for this
$T_1 = 21°C = 294\text{K}$ $T_2 = 0°C = 273\text{K}$

The problem involves pressure, volume, and temperature changes, so the combined gas law may be used.

$$\frac{P_1 V_1}{T_1} = \frac{P_2 V_2}{T_2}$$

rearranging to solve for V_2: $\quad V_2 = \frac{P_1 V_1 T_2}{T_1 P_2}$

solve: $\quad V_2 = \frac{(0.75\,\text{atm})(50.\,\text{mL})(273\text{K})}{(249\text{K})(1\,\text{atm})} = 41.11\,\text{mL} = 41\,\text{mL}$

6.   (1) $1.25\,\dfrac{\text{g}}{\text{L}}$

The molar mass of CO is 12.01 g/mol + 16.00 g/mol = $28.01\,\dfrac{\text{g}}{\text{mol}}$

$$d = \left(\frac{28.01\,\text{g}}{\cancel{\text{mol}}}\right)\left(\frac{1\,\cancel{\text{mol}}}{22.4\,\text{L}}\right) = 1.25\,\frac{\text{g}}{\text{L}}$$

(2)   82.6 L HCN

Knowns:

$\qquad\qquad$ density at STP, d = 1.21 g/L   V at STP = ? (solve for this)

$\qquad\qquad$ mass, m = 100. g

Equation $d = \dfrac{m}{V}$

Rearrange to solve for V (multiply each side of the equation by V, and divide each side by d):

$$V = \frac{m}{d}$$

Calculate: $V = \dfrac{100.\ \cancel{g}}{1.21\ \cancel{g}/L} = 82.6\ L$

(3)   156 mL

Since the acetylene gas was collected over water, we have to subtract the partial pressure of water vapor at 21°C to find the partial pressure of acetylene alone. Then, we can find the volume at STP.

$$P_{total} = P_{C_2H_2} + P_{H_2O}$$

$$750\ torr = P_{C_2H_2} + 18.6\ torr\ (Appendix\ II)$$

$$P_{C_2H_2} = 750\ torr - 18.6\ torr = 731\ torr$$

Establish a table of knowns

$P_1 = 731.4\ torr$	$P_2 = 760.\ torr\ (standard)$
$V_1 = 175\ mL$	$V_2 = x\ mL$
$T_1 = 21°C = 294\ K$	$T_2 = 0°C = 273\ K\ (standard)$

The problem involves pressure, volume, and temperature changes, so the combined gas law may be used.

$$\frac{P_1 V_1}{T_1} = \frac{P_2 V_2}{T_2}$$

rearranging to solve for V_2: $V_2 = \dfrac{P_1 V_1 T_2}{T_1 P_2}$

pressure increases and temperature decreases, both of which would lead to a reduction in volume

solve: $V_2 = \dfrac{(731\ \cancel{torr})(175\ mL)(273\ K)}{(294\ K)(760\ \cancel{torr})} = 156\ mL$

7. 1.5×10^2 L (2 significant figures)

Knowns:

PV = nRT (ideal gas equation)   V = ? (solve for this) $V = \dfrac{nRT}{P}$
mass, m = 15 g   gas is H_2
T = 25°C = 298K   P = 1.2 atm
R = 0.0821 L.atm/mol.K

Plan: To be able to use PV = nRT, we need to know n (moles)
         g H_2 → n H_2 → ideal gas equation to solve for V

solve: moles, $n = \dfrac{mass}{molar\ mass} = (15\ \cancel{g})\left(\dfrac{1\ mol\ H_2}{2.0\ \cancel{g}}\right) = 7.5\ mol\ H_2$

$$V = \frac{(7.5\ \cancel{mol})\left(0.0821\ \dfrac{L\cdot \cancel{atm}}{\cancel{mol}\cdot \cancel{K}}\right)(298\ \cancel{K})}{1.2\ \cancel{atm}}$$

$$= 1.5 \times 10^2\ L$$

8.   52 L

The number of moles of starting substance is 7.0 moles NO_2.
Calculate the moles of NO

$$(7.0\ \cancel{mol\ NO_2})\left(\frac{1\ mol\ NO}{3\ \cancel{mol\ NO_2}}\right) = 2.3\ mol\ NO$$

Convert moles of NO to L of NO. The moles of a gas at STP are converted to L by multiplying by the molar volume, 22.4 L per mole:

$$(2.3\ \cancel{\text{mol}}\ \text{NO})\left(\frac{22.4\ \text{L}}{\cancel{\text{mol}}}\right) = 52\ \text{L NO}$$

9. 134 L

This is a volume-volume calculation and is an application of Avogadro's Law. Also, we need to consider whether there is a limiting reactant. First, determine how many moles of NO(*g*) can be produced from each reactant.

$$(4.00\ \cancel{\text{mol}}\ \text{N}_2)\left(\frac{2\ \text{mol NO}}{1\ \cancel{\text{mol}}\ \text{N}_2}\right) = 8.00\ \text{mol NO}$$

$$(3.00\ \cancel{\text{mol}}\ \text{O}_2)\left(\frac{2\ \text{mol NO}}{1\ \cancel{\text{mol}}\ \text{O}_2}\right) = 6.00\ \text{mol NO}$$

Therefore, O_2 is the limiting reactant and 6.00 mol NO will be produced in the reaction with 1.00 mol of N_2 left over.

The volume of 6.00 mol of NO at STP will be

$$(6.00\ \cancel{\text{mol}})\left(22.4\frac{\text{L}}{\cancel{\text{mol}}}\right) = 134\ \text{L}$$

10. 1.80×10^2 moles O_2 and 5.76×10^3 g O_2

We need to use the ideal gas equation to solve the problem. The data tabulated looks like this:

$$\begin{aligned}
P &= (1500\ \text{psi}) = (1500\ \cancel{\text{psi}})\left(0.06805\frac{\text{atm}}{\cancel{\text{psi}}}\right) = 1.0 \times 10^2\ \text{atm} \\
V &= 43.8\ \text{L} \\
R &= 0.0821\frac{\text{L}\cdot\text{atm}}{\text{mol}\cdot\text{K}} \\
T &= 72°\text{F} = 22°\text{C} = 295\ \text{K} \\
PV &= nRT \\
n &= \frac{PV}{RT} = \frac{(1.0 \times 10^2\ \cancel{\text{atm}})(43.8\ \cancel{\text{L}})}{\left(0.0821\frac{\cancel{\text{L}}\cdot\cancel{\text{atm}}}{\text{mol}\cdot\cancel{\text{K}}}\right)(295\ \cancel{\text{K}})} = 1.80 \times 10^2\ \text{mol O}_2
\end{aligned}$$

The mass of oxygen equals

$$(1.80 \times 10^2\ \cancel{\text{mol}}\ \text{O}_2)\left(32.00\frac{\text{g}}{\cancel{\text{mol}}}\right) = 5.76 \times 10^3\ \text{g}$$

11. 77.7 g/mol, AsH_3

$$\begin{aligned}
&\text{mass, g} = 3.48\ \text{g} \\
&\text{P} = 740.\ \text{torr} = 0.974\ \text{atm} \\
&\text{T} = 21°\text{C} = 294\ \text{K} \\
&\text{V} = 1.11\ \text{L}
\end{aligned}$$

Using $PV = nRT$ and moles, $n = mass\ in\ grams/molar\ mass$,

we can combine and rearrange the equation to get molar mass $(\text{M}) = \frac{\text{gRT}}{\text{PV}}$

Substituting into the equation:

$$M = \frac{(3.48\text{g})(0.0821\ \cancel{L} \cdot \cancel{atm})(294\ \cancel{K})}{(0.974\ \cancel{atm})(1.11\ \cancel{L})(\text{mol} \cdot \cancel{K})}$$

$$= 77.7\ \text{g/mol}$$

Subtracting the mass of the three hydrogens ($77.7 - 3(1.01) = 74.7$ g/mol) gives the mass of the metal: As.

CHAPTER THIRTEEN

Properties of Liquids

SELECTED CONCEPTS REVISITED

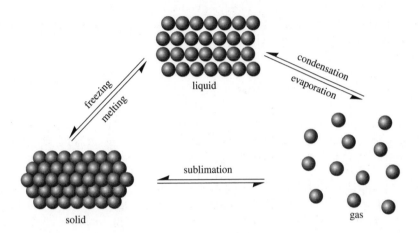

The molecules in solids have low average kinetic energy. As the temperature is raised the molecules gain kinetic energy raising the average kinetic energy of the molecules in the sample. If the molecules gain enough energy, they eventually break enough of the <u>inter</u>molecular attractions to become a liquid and if all the attractions are overcome, a gas.

The vapor pressure of a liquid is the pressure exerted by the vapor above a liquid when the rate of evaporation is equal to the rate of condensation. Using water as an example, suppose you have a sample of water in a closed container. Some of the water molecules may possess enough kinetic energy to break free of the other molecules and achieve the gaseous state (that is, some molecules will become water vapor). The water vapor (a gas) will exert pressure on the walls of the container. Eventually some of the molecules of the water vapor may lose kinetic energy and condense to the liquid state. At some point in time this process will achieve equilibrium, that is the rate of the two processes (vaporization and condensation) will be equal. The vapor pressure is the pressure exerted by the vapor when a liquid is in equilibrium with its vapor. Because the average kinetic energy of liquid is dependent on its temperature, the vapor pressure of a liquid is also temperature dependent.

Liquids with a higher vapor pressure evaporate faster and the liquid is considered to be more volatile. Liquids with higher vapor pressures tend to have weaker and/or fewer intermolecular attractions between liquid molecules and therefore do not need as much energy to overcome these attractions. Note that you need not memorize this; it is better to understand vapor pressure and you can then rationalize the trend. If you have two liquids under similar conditions and assume there is no vapor initially present, then when both have reached the liquid-gas equilibrium for the same temperature, the only way one liquid can have a higher vapor pressure is if more molecules are in the gaseous state at equilibrium. So higher vapor pressure means more gas molecules present when rate of evaporation equals rate of condensation, therefore it is probably easier to vaporize that liquid, so there are weaker intermolecular attractions in the liquid state.

The boiling point of a liquid is the temperature at which its vapor pressure is equal to the external (atmospheric) pressure. Therefore changing the external pressure can alter the actual boiling point of a liquid. If the external pressure is raised, that means the boiling point of the liquid is also raised. Here is a possible rationalization:

Higher external pressure means the liquid needs to have a higher vapor pressure before it will boil. A higher vapor pressure means more molecules need to have escaped from the liquid to get into the gas state. To get more molecules into the gas state, the liquid needs to have a higher average kinetic energy. Higher temperatures cause higher kinetic energies of molecules. So higher external pressure causes a higher boiling point.

Alternatively, low external pressure means a lower vapor pressure is necessary therefore fewer molecules in the vapor state and so the liquid can boil at a lower temperature.

Water, H_2O, is a polar covalent bent molecule, with the O having a $\delta-$ and each H having a $\delta+$ charge (δ means partial). Liquid water exhibits an intermolecular attraction called a **hydrogen bond**. A hydrogen bond is an electrostatic attraction between the $\delta-$ O of one water molecule and the $\delta+$ H on another molecule. A hydrogen bond is simply such a good dipole-dipole interaction that it warrants its own name.

A hydrogen bond is formed between a really good $\delta-$ and a really good $\delta+$. How do we know what constitutes a really good $\delta-$ or $\delta+$? Only the three most electronegative elements are able to have a good enough $\delta-$ to participate in hydrogen bonding; thus only $\delta-$ which reside on F, O, and/or N. Only H can achieve a good enough $\delta+$ to participate in hydrogen bonding but not all! H's. To get a good $\delta+$ H, that H must be bonded to an atom which really shares the electrons unequally, that is the H must be attached to a very electronegative atom. So only H's attached to an F, O or N have a good enough $\delta+$ to participate in hydrogen bonding.

COMMON PITFALLS TO AVOID

Do not confuse boiling point with vapor pressure. The normal boiling point is the temperature at which the vapor pressure is equal to the atmospheric pressure (1 atm) above the liquid. The vapor pressure is the pressure exerted by the vapor above the liquid when the liquid and its vapor are at equilibrium. The vapor pressure is temperature dependent.

Remember that it takes energy to actually carry out a phase change. That is, in going from a solid to a liquid, there will be some time when the solid is absorbing heat but there is no visible change in temperature. The heat being absorbed during this time is being used to break some of the intermolecular attractions and not into raising the temperature. A similar situation occurs in going from a liquid to a gas. So if you are trying to determine how much heat is absorbed when ice at $-10\,°C$ to water at $50\,°C$ do not forget to include the heat absorbed during the phase change when there is no temperature change.

A hydrogen bond is not a covalent bond! Only H's attached to F, O, or N can participate in hydrogen bonding but please remember that those covalent bonds are not hydrogen bonds.

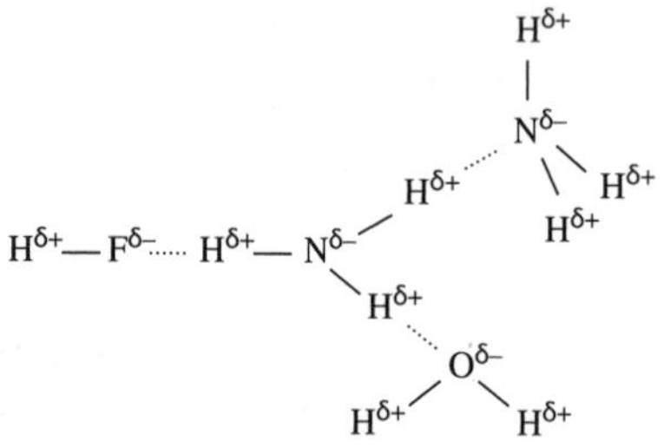

dotted lines represent hydrogen bonds

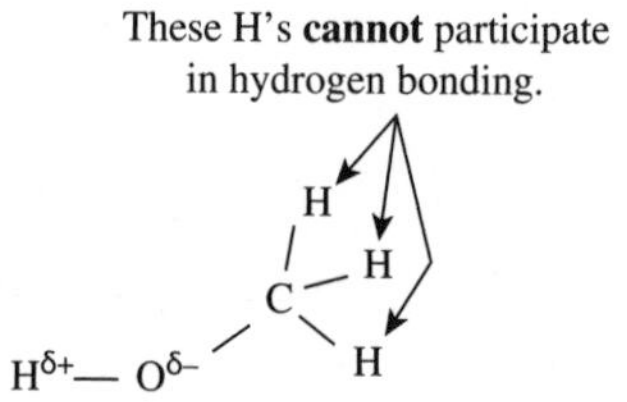

These H's **cannot** participate in hydrogen bonding.

solid lines represent covalent bonds

(Note, it is unlikely you will have a flask containing these molecules together. They are shown here to illustrate hydrogen bonding.)

When water boils, H_2O molecules remain as H_2O molecules. The hydrogen bonds between H_2O molecules are broken but not the bonds in the H_2O molecule itself.

RECAP SECTION

Chapter 13 introduces many new terms you need to be familiar with and also focused significantly on water. Why liquids tend to form drops, the process of evaporation and its relationship to vapor pressure are all topics explored in this chapter. Graphs of temperature vs vapor pressure, the definitions of boiling point and melting point, heating curves and using them to aid in the calculation of the amount of energy involved are also covered. The types of intermolecular forces introduced here will crop up time and again in future chapters and understanding their importance will help you predict relative boiling points. This chapter also takes a look at hydrates, what they are, how their formulas are written, and how to calculate the percent water in a hydrate. The structure and sources of water are also explored.

SELF-EVALUATION SECTION

1. Many of the important characteristics of water are related to its physical properties. For example, the amount of heat required to change 1 gram of a solid into a liquid, called the (1) _____, is an unusually large amount for water, as is the amount of heat required to change 1 gram of liquid at its normal boiling point into a gas called the (2) _____. The temperature at which ice begins to change into the liquid state is called the (3) _____ and the temperature at which the vapor pressure of water equals the atmospheric pressure of 1 atm is called the (4) _____.

2. Water, which is the most common chemical substance around us, has been the subject of many experiments over the years. It is a simple molecule, H_2O, and yet its properties suggest that water is a very large molecule. We want to analyze how the structure of water influences such properties as boiling point and melting point.

    Fill in the blank space or circle the appropriate response.

    A single water molecule consists of (1) _____ H atom(s) and (2) _____ O atom(s). The oxygen atom is the middle atom joined to each H atom by a(n) (3) ionic/covalent bond. The molecule is (4) straight end to end/bent in the middle with a bond angle of 105° between the two covalent bonds.

    Oxygen is a very (5) electropositive/electronegative element and, as a result, the two covalent OH bonds are (6) nonpolar/polar. The bend in the molecule in conjunction with the polar covalent bonds make water a (7) nonpolar/polar molecule. The oxygen atom carries a partial (8) _____ charge, and each hydrogen atom carries a partial (9) _____ charge.

    Since each water molecule has the same unequal charge distribution, there will be an attraction between molecules. The oxygen side of the molecule will be attracted to the (10) _____ atom of another water molecule through a weak electrostatic bond. This type of attraction between an H atom and a highly electronegative atom is called a (11) _____ bond. The weak ionic association between water molecules produces the effect of water behaving as a large molecule. When we examine the physical properties of water, it is clear that water does not fit the expected pattern. The melting point and normal boiling point are (12) higher/lower than expected, as are the heat of fusion and heat of vaporization. Water acts though it were a large bulky molecule rather than a small one with a molar mass of 18.02 g/mol. The bent structure and resulting polarity have a marked influence on the physical properties of water.

3. Fill in the blank space or circle the appropriate response.

    All substances in the liquid state are in the process of vaporizing. Some chemicals, such as acetone and ether, evaporate quickly, whereas others, such as mercury, do so slowly. The process of molecules going from the liquid state to the gas state is called (1) _____. In any sample of a liquid, (2) repulsive/ attractive forces exist that must be overcome before a molecule can escape the liquid state. Even though the temperature of the liquid is uniform, not all of the molecules possess the same (3) kinetic/potential energy. Since the masses are constant, this suggests that the velocities of the molecules are (4) the same/different. Therefore, molecules at the surface of the liquid, which are moving faster than their neighbors, are able to overcome the attractive forces and escape to the gas state. After these molecules with greater kinetic energy leave the liquid state, the average kinetic energy of the remaining molecules in the liquid state is (5) lowered/ raised. Therefore, the temperature (6) rises/drops, and we find that vaporization is a (7) warming/cooling process, which we know to be true from everyday experience. If the temperature drops as the average kinetic energy of the system goes down, then where does the heat energy come from to vaporize all the water from an open dish? (8) _____

4. (1) In terms of pressure, when will a liquid boil?
    (2) What are the values of atmospheric pressure, in torr, and in atm?
    (3) What can cause changes in atmospheric pressure?
    (4) At the top of a mountain, would you expect the boiling point of a liquid to increase or decrease? Explain briefly in terms of pressure.

5. On the following cooling curve, a gaseous substance at point A is cooled until it reaches point E. Identify the various stages along the curve as requested.

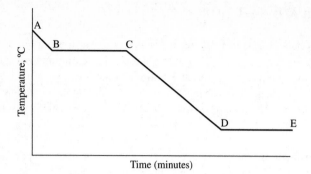

Time (minutes)

Identification of Process or Condition of State

(1) Point B   _____

(2) Line BC   _____

(3) Line CD   _____

(4) Point D   _____

(5) Line DE   _____

6. Identify each of the salt formulas as anhydrous or hydrates. Name them.

(1) $CaSO_4$ _____

(2) $CoCl_2 \cdot 6H_2O$ _____

(3) $NaC_2H_3O_2 \cdot 3H_2O$ _____

(4) K_2S _____

(5) Na_3PO_4 _____

7. For each of the following processes, sketch and label a simple heating or cooling curve and write a solution map you could use if you had to calculate the energy change.

(1) liquid lauric acid at 55°C is cooled to 22°C (f.p. is 44°C).
(2) ice at −5°C is warmed to steam at 100°C.
(3) ethyl alcohol at 54.3°C was heated to 98.5°C (b.p. is 78.4°C)

8. Two 35.0 samples of gaseous ethanol (ethyl alcohol), C_2H_5OH, at 78.4°C are cooled until they are solidified. Given the following information

Ethanol: melting point −112°C
                                         boiling point 78.4°C
                                         heat of fusion 29.4 cal/g
                                         heat of vaporization 204.3 cal/g
                                         specific heat, Cp 2.49 joule/g°C

1 Joule = 0.239 calories

(1) Draw a cooling curve to illustrate the process, clearly labeling relevant temperatures and physical states.
(2) How many calories of heat energy are lost during this process?

Draw your cooling curve here.

Do your calculations here.

9. Refer to Fig 13.7 in your textbook. Would you expect the vapor pressure curve for ethane (CH_3CH_3) to be to the far left, the far right, or between two of the existing curves? Briefly explain.

10. For each of the following compounds, indicate the strongest type of intermolecular forces you would find in a pure sample of the compound.
(a) CH_3NH_2 (b) $CH_3CH_2CH_3$ (c) $H_3C-O-CH_3$
(d) CH_3CH_2F (e) CH_3CH_2OH (f) CH_3Cl

11. Excluding (a), which of the compounds in the previous example would you expect to have the highest normal boiling point and which the lowest? Explain briefly.

12. Indicate clearly which hydrogen(s), if any, in each of the following molecules are able to participate in hydrogen bonding with a water molecule.
    (1) CH_3CH_2OH　　　　(2) CH_3F　　(3) CH_3OCH_3　　(4) H_3CCH_2COOH
    (5) NH_3　　　　　　　(6) CH_3NHCH_3

Challenge Problems

13. An ice cube at 0°C has a mass of 7.25 g.

    (1) In order for the ice cube to be heated up enough to vaporize, which stages of the heating curve must it pass through? Draw and label the heating curve showing this process.
    (2) Use your answer to help you calculate the heat (in joules) needed to convert the ice cube at 0°C to steam at 100°C.

    The heat of fusion is 335 J/g. The heat of vaporization is 2.26 kJ/g. The specific heat is 4.184 J/g°C.

    Draw your curve here.

    Do your calculations here.

ANSWERS TO QUESTIONS

1. (1) heat of fusion　　　(2) heat of vaporization　(3) melting point
   (4) normal boiling point　(5) basic anhydrides　(6) acid anhydrides
   (7) hydrates

2. (1) 2　　　　　　　　(2) 1　　　　　　　　(3) covalent
   (4) bent in the middle　(5) electronegative　(6) polar
   (7) polar　　　　　　(8) negative　　　　(9) positive
   (10) hydrogen　　　　(11) hydrogen　　　　(12) higher

3. (1) vaporization　　　(2) attractive　　　　(3) kinetic
   (4) different　　　　　(5) lowered　　　　　(6) drops
   (7) cooling　　　　　(8) the surroundings

4. (1) A liquid will boil when its vapor pressure is equal to the atmospheric pressure.
   (2) 760 torr, or 1 atm

(3)   Atmospheric pressure can vary with the weather, and with changes in elevation.

(4)   The boiling point would decrease as the atmospheric pressure is generally lower at higher elevations, therefore the vapor pressure will equal the lower atmospheric pressure at lower temperatures.

5.   (1)   melting point   (2)   solid in equilibrium with liquid

        (3)   all-liquid state, temperature begins to rise   (4)   boiling point

        (5)   boiling liquid in equilibrium with gas

6.   (1)   anhydrous calcium sulfate   (2)   hydrate : cobalt (II) chloride hexahydrate   (3)   hydrate : sodium acetate trihydrate

        (4)   anhydrous potassium sulfide   (5)   anhydrous sodium phosphate

7.   (1)

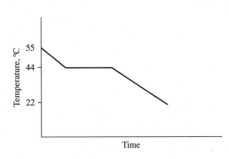

    (2)

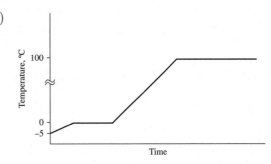

Solution map: liquid 55°C → 44°C
                 freezing liq. 44°C → solid 44°C
                 solid 44°C → 22°C

Solution map: ice −5°C → 0°C
                 melting solid 0°C → liq. 0°C
                 liquid 0°C → 100°C
                 vaporizing liq. 100°C → gas 100°C

    (3)

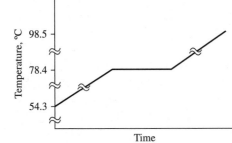

Solution map: liquid 54.3°C → 78.4°C
                 vaporizing liq. 78.4°C → gas 78.4°C
                 gas 78.4°C → 98.5°C

8.   (1)

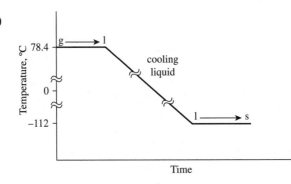

    (2)   Knowns are listed in the question.

           Plan: gaseous ethanol at 78.4°C → liquid ethanol at 78.4°C   use heat of vaporization
                    liquid ethanol at 78.4°C → liquid ethanol at −112°C   use specific heat
                    liquid ethanol at −112°C → solid ethanol at −112°C   use heat of fusion

$$\text{condensing ethanol (energy loss)} = \text{(mass)(heat of vaporization)}$$

$$= (35.0 \text{ g})\left(\frac{204.3 \text{ cal}}{\text{g}}\right) = 7150.5 \text{ cal} = 7150 \text{ cal}$$

$$\text{cooling liquid ethanol} = \text{(mass) (specific heat) (change in temp)}$$

$$= (35.0 \text{ g})\left(\frac{2.49 \text{ joule}}{\text{g} \, ^\circ\text{C}}\right)\left(\frac{0.239 \text{ cal}}{1 \text{ joule}}\right)(90.4 \, ^\circ\text{C})$$

$$= 1882.9 \text{ cal} = 1880 \text{ cal}$$

$$\text{solidifying ethanol} = \text{(mass)(heat of fusion)}$$

$$= (35.0 \text{ g})\left(\frac{24.9 \text{ cal}}{\text{g}}\right) = 871.5 \text{ cal} = 872 \text{ cal}$$

$$\text{total energy lost} = 7150 + 1880 + 872 = 9902 \text{ cal}$$

$$= 9.90 \times 103 \text{ cal} = 9.90 \text{ kcal}$$

Note: pay attention to significant figures!

9. The v.p. curve should be to the far left, as with only weak London dispersion forces, the boiling point of ethane should be lower than that of ethyl ether.

10. (a) Hydrogen bonding
    (b) London dispersion – no polar bonds
    (c) Dipole-dipole – has polar C—O bonds; no OH bonds, so no hydrogen bonding
    (d) Dipole-dipole – has polar C—F bond; no H—F bonds, so no good H to take part in hydrogen bonding
    (e) Hydrogen bonding
    (f) Dipole-dipole – has polar C—Cl bond

11. All of the compounds, b-f, have relatively similar molar masses, so CH_3CH_2OH would have the highest boiling point because it has the strongest intermolecular forces. With only London dispersion forces, $CH_3CH_2CH_3$ should have the lowest boiling point. The stronger the intermolecular forces in the liquid state, the more energy is needed to break them in order to achieve the gaseous state, and thus the higher the boiling point.

12. The hydrogens that can participate in hydrogen bonding are highlighted in bold; only those attached <u>directly</u> to an O, N, or F.

    (1) $CH_3CH_2\mathbf{OH}$ (2) none   (3) none   (4) $H_3CCH_2COO\mathbf{H}$

    (5) $\mathbf{NH_3}$ (6) $CH_3\mathbf{N}HCH_3$

13. (1)

    Plan: ice at 0°C → water at 0°C   use heat of fusion
           water at 0°C → water at 100°C   use specific heat
           water at 100°C → steam at 100°C   use heat of vaporization

(2) To melt the ice cube requires 335 J for each gram. Therefore,

$$\left(335\,\frac{J}{g}\right)7.25\ g = 2430\ J$$

We now have water at 0°C and will need 4.184 J per gram of water per degree to raise the water temperature to boiling. Therefore,

$$(7.25\ g)(100\,°C)\left(4.184\,\frac{J}{g\cdot°C}\right) = 3030\ J$$

Notice how the units cancel to give us units of heat energy – joules. Now we must use 2.26 kJ for each gram of hot water to convert it into steam.

$$(7.25\ g)\left(2.26\,\frac{kJ}{g}\right) = 16.4\ kJ$$

The last step is to add up the individual values after converting to kJ.

$$2.43\ kJ + 3.03\ kJ + 16.4\ kJ = 21.9\ kJ$$

(If you waited until the end of the calculation to round off, the answer is 21.865)

CHAPTER FOURTEEN

Solutions

SELECTED CONCEPTS REVISITED

True liquid solutions are homogeneous mixtures and often visibly transparent. The solute is molecular or ionic in size, does not settle out, is uniformly distributed, and can often be easily separated from the solvent by physical means.

Two nonpolar substances easily intermingle to form a solution. The dissolution of ionic compounds in water is somewhat more complicated. Water is very polar and can attract positively and negatively charged ions. These attractions are generally strong enough to allow the ions to move away from each other with the net result being that the ions are distributed in the solution. Each ion is surrounded by water molecules whose opposite pole is facing inwards (that is, the hydrogens of water face in towards a negative charge, while the oxygen of water faces inwards to a positive charge).

The effect of temperature on the solubility of a solid in a liquid is somewhat unpredictable. For some solids, their solubility decreases but for most solids the solubility increases as temperature increases. The temperature effect on the solubility of gases is more predictable. In general, as the temperature of the solution increases, the solubility of the gas in the solution decreases.

A change in pressure tends to significantly affect the solubility of a gas in a liquid. The solubility of the gas tends to rise proportionately to an increase in pressure.

Qualitative descriptions of aqueous solutions include the following terms:

Dilute – only a small amount of solute is present *relative* to the volume of the solution.
Concentrated – a large amount of solute is present *relative* to the volume of the solution.
Unsaturated – the solution is capable of dissolving more solute.
Saturated – the maximum amount of solute relative to the volume of the solution has been dissolved. Any excess solute added will not dissolve. The rate of solute dissolving to the rate of solute crystallizing out of solution is equal (at equilibrium).
Supersaturated – more solute than can normally be dissolved in solution has dissolved. Note that this type of solution must be prepared carefully and the excess solute is easily crystallized out (often triggered by something as simple as moving the container.)

Quantitative descriptions of the concentration of aqueous solutions include the following units: Mass percent, ppm, mass/volume %, volume %, molality, and perhaps the most commonly used term, molarity.

The concentration of a solution is most often described using units of molarity (M).
    M = moles solute/liters of solution.

In a dilution, solvent is added; the volume of the solvent increases and therefore the volume of the solution also changes. Since the moles of solute remain constant but the volume of the solution increases, the molarity (or concentration) of the solution decreases.

Colligative properties are the properties of a solution that depend on the number of particles in solution. For example, consider the effect of an increased number of nonvolatile particles on the boiling point. The more nonvolatile particles in solution, the lower the vapor pressure which in turn means that more heat needs to be applied to the solution to get more molecules into the vapor phase before the vapor pressure equals the atmospheric pressure which leads to a higher boiling point.

COMMON PITFALLS TO AVOID

Not all liquids are solutions, not all solutions are liquids (although you will deal mainly with liquid aqueous solutions) and not all mixtures are solutions.

Saturated is not the same as concentrated. A solution could be saturated and still be dilute. For example, if a solute is not very soluble in water, then even when as much solute as possible has been dissolved, relative to the volume of water present the amount of solute is small and therefore even though the solution is saturated, it may still be considered dilute.

Molarity and molality may sound similar but these two units of concentration are not the same. Molarity is generally used to describe most aqueous solutions and has units of mol/L. Molality is more commonly used in freezing point depression or boiling point elevation calculations and has units of mol/kg.

Realize that $M_1L_1 = M_2L_2$ is a shortcut for a dilution problem. You should only use shortcuts that you understand. Molarity $\times$ volume (in L) gives moles of solute. Since the moles of solute do not change during a dilution, the molarity $\times$ volume before dilution is equal to the molarity $\times$ volume after the dilution. Gaining a good understanding of the concept behind $M_1L_1 = M_2L_2$ will help you use the formula only at the appropriate times, when moles of solute are not changing.

Use caution when using or interpreting equation symbols. For example, M represents molarity, m can represent mass or molality, and mol (not m) represents moles. It should be noted that molality is often an italicized m in texts but in hand-written formula, it is difficult to distinguish from the mass, m. Be sure to use the appropriate quantity for your calculations. K's (and k's) are often also used in multiple situations.

RECAP SECTION

Chapter 14 explores solutions, both quantitatively and qualitatively. The properties of a true solution, solubility and the factors which affect it and the rate of dissolution are discussed. Quantitative problems involving mass percent, volume percent, molarity and dilution are used in many scientific fields, not just chemistry, and you should understand and be comfortable with these types of calculations. Colligative properties, where solutions are generally quantified using molality, affect freezing points and boiling points, leading to freezing point depression, and boiling point elevation, of various solutions. Osmosis and osmotic pressure are defined and their importance and use in nature is outlined. The preparation and calculations involving solutions is a skill that bench chemists and lab technicians use every day and is critical in fields such as nursing, agriculture, and food technology. There are many good review exercises both in this book and at the end of the chapter in the text. Do any problems assigned and then try a few more to sharpen your problem-solving ability.

SELF-EVALUATION SECTION

1. In the following statements, identify the (a) solution, (b) solute, and (c) solvent.

    (1) Air is composed primarily of two gases – oxygen and nitrogen. Air is approximately 79% nitrogen and 21% oxygen.

       a. _____   b. _____   c. _____

    (2) In order for certain species of fish to thrive in lakes and streams, the dissolved oxygen content of the water has to be 0.0005% or greater.

       a. _____   b. _____   c. _____

    (3) A 0.1 M iodine solution was prepared by dissolving crystals of iodine in carbon tetrachloride (CCl_4).

       a. _____   b. _____   c. _____

    (4) Household bleach is a 5% solution of sodium hypochlorite (NaClO) in water.

       a. _____   b. _____   c. _____

2. Solubility of salts in water.

All nitrates, (1) _____, and (2) _____, and ammonium salts are soluble in water.

All (3) _____, bromides, and (4) _____ are soluble in water, except those salts of (5) _____, (6) _____, and lead (II).

All carbonates and (7) _____, and are insoluble in water except those of Na^+, K^+, and (8) _____.

All (9) _____ are insoluble in water except those of (10) _____, Group 1A and Group 2A metals.

Place an *s* (for soluble) or *i* (for insoluble) next to each of these compounds, if H_2O is the solvent.

(11)	Na_2S	_____	(21)	$Al(NO_3)_3 \cdot 9H_2O$	_____
(12)	Na_2CO_3	_____	(22)	K_3PO_4	_____
(13)	BaI_2	_____	(23)	$(NH_4)_2CO_3$	_____
(14)	NH_4NO_3	_____	(24)	KCl	_____
(15)	$NiCO_3$	_____	(25)	$Mg_3(PO_4)_2$	_____
(16)	$LiI \cdot 3H_2O$	_____	(26)	$PbCl_2$	_____
(17)	$Sn(NO_3)_4$	_____	(27)	$HgBr$	_____
(18)	Fe_2S_3	_____	(28)	$ZrCl_4$	_____
(19)	$FePO_4$	_____	(29)	Ag_2S	_____
(20)	AgI	_____	(30)	$CaBr_2$	_____

3. Solubilities of salts in g/100 g H_2O.

$CaBr_2$	125 g at 0°C	$(NH_4)_2CO_3$	100 g at 15°C
NH_4NO_3	118 g at 0°C	$NaC_2H_3O_2$	119 g at 0°C
Na_2S	15.4 g at 10°C	$LiI \cdot 3H_2O$	151 g at 0°C
$Na_2S_2O_3$	50 g at 20°C	KCl	34.7 g at 20°C
$MgSO_4$	26 g at 0°C	Na_2CO_3	7.1 g at 0°C
$Al(NO_3)_2$	63.7 g at 25°C	K_3PO_4	90 g at 20°C

Identify the following solutions as saturated, unsaturated, or supersaturated. (All solutes are in 100 g H_2O.)

(1) 5 g of $CaBr_2$ at 0°C

(2) 100 g of NH_4NO_3 at 0°C

(3) 15.4 g of Na_2S at 10°C

(4) 55 g of $Na_2S_2O_3$ at 20°C

(5) 4 g of $MgSO_4$ at 0°C

(6) 51 g of $Al(NO_3)_3$ at 25°C

(7) 75 g of $(NH_4)_2CO_3$ at 15°C

(8) 125 g of $NaC_2H_3O_2$ at 0°C

(9) 38 g of $LiI \cdot 3H_2O$ at 0°C

(10) 34.7 g of KCl at 20°C

(11) 6.5 g of Na_2CO_3 at 0°C

(12) 12 g of K_3PO_4 at 20°C

4. Fill in the blank space or circle the appropriate response.

Water molecules, which are very (1) polar/nonpolar, are (2) attracted/repulsed by other polar or ionic molecules or ions. Water molecules weaken the ionic forces that hold ions such as Na^+ and Cl^- together. Then the ions are pulled apart by the interaction with water as the water molecules surround the ions. The charged ions such as Na^+ and Cl^- then diffuse slowly away from the mass of undissolved salt as hydrated ions.

With most solid chemicals, an increase in temperature means an (3) _____ in solubility. However, a gaseous chemical always (4) _____ in solubility as the temperature increases. A temperature increase means that the kinetic energy of the gas molecules (5) _____ and their ability to associate with the solvent molecules (6) _____.

Pressure changes don't affect the solubility of solids greatly but gases show marked changes. The solubility of a gas is (7) inversely/directly proportional to the pressure of the gas above the liquid. Double the pressure means the solubility will (8) _____. Carbonated beverages are a good example.

5. The solubility of salts can vary considerably with temperature. Although the solubility of most substances increase with increasing temperature, some decrease. Using Figure 14.4 in your text, list the following solutes in order of increasing solubility.

KCl, $CuSO_4$, HCl, Li_2SO_4, KNO_3 (all at 1 atm)

(1) At 10°C _____, _____, _____, _____, _____ (most soluble)
(2) At 50°C _____, _____, _____, _____, _____ (most soluble)
(3) If 20 g of each solute above was mixed into 50 g of water, in which flasks would you expect to see undissolved solid; at 10°C _____? at 50°C _____?

6. List the four factors that influence the rate or speed at which a solid solute dissolves.

7. (1) What masses of sodium chloride (NaCl) and water are needed to make 525.0 g of 0.1000% salt solution?

Do your calculations here.

(2) A 1.5 L bottle of cooking sherry contains 18% alcohol by volume. How many mL of alcohol are in the bottle?

Do your calculations here.

(3) A 250. mL solution contains 3.8 moles of H_2SO_4. What is the molarity of the solution?

Do your calculations here.

(4) If 2.6 moles of KCl is used to make a 4.0 L batch of solution, calculate the molarity of the solution.

Do your calculations here.

(5) How would you prepare 1500 g of a 2.000% sucrose (sugar) mass percent solution to be used for intravenous feeding in a hospital?

Do your calculations here.

8. Calculate how many grams of chemical would be required to prepare the following solutions.

Atomic Masses, amu

(1) 600. mL of 3.00 M NaOH   Na = 22.99   F = 19.00
(2) 4.00 L of 1.00 M $(NH_4)_2SO_4$ S = 32.07   Ca = 40.08
(3) 420 mL of 0.70 M $CaSO_4$ O = 16.00   C = 12.01
(4) 250 mL of 0.94 M $Ca(NO_3)_2$ N = 14.01   H = 1.008
(5) 0.50 L of 0.10 M NaF

Use the relationship

$$M = \frac{g \text{ of solute}}{\text{molar mass solute} \times L \text{ of solution}}$$

or solve by dimensional analysis.

Do your calculations here.

9. More experience in working with molarities of solutions.

(1) What volume of 1.25 M $K_2Cr_2O_7$ can be prepared from 130. g of $K_2Cr_2O_7$?

Atomic Masses, amu

K = 39.10
Cr = 52.00
O = 16.00

Do your calculations here.

(2) What volume of 0.400 M $NaHCO_3$ can be prepared from 48.0 g of $NaHCO_3$?

Atomic Masses, amu

$Na = 22.99$
$H = 1.008$
$O = 16.00$
$C = 12.01$

Do your calculations here.

(3) Calculate the number of moles of hydrochloric acid in 2.5 L of 0.124 M HCl.

Do your calculations here.

10. Calculate the number of moles of solute contained in the following solutions:

(1) 250 mL of 0.22 M $K_2Cr_2O_7$
(2) 1500 mL of 1.4 M Na_2CO_3
(3) 3.15 L of 0.75 M $HClO_4$
(4) 857 mL of 0.66 M $CuSO_4 \cdot 5H_2O$

Do your calculations here.

11. Using the relationship for dilutions, calculate how much (mL) concentrated reagent is necessary for the following solutions.

$$(volume_1)(M_1) = (volume_2)(M_2)$$

(1) 15 M NH_4OH to prepare 2.5 L of 5.0 M NH_4OH
(2) 14.6 M H_3PO_4 to prepare 150. mL of 0.30 M H_3PO_4
(3) 12 M HCl to prepare 500. mL of 0.25 M HCl
(4) 18 M H_2SO_4 to prepare 250. mL of 1.25 M H_2SO_4

Do your calculations here.

12. Water from the Great Salt Lake can have a salt concentration as high as 3.42 M (expressed as NaCl). High mountain spring water has a very low salt concentration, 0.00171 M. Potable, or drinkable, water has a maximum recommended level of 0.0171 M or 10 times that of spring water. What is the maximum volume of drinkable water that can be prepared from 500 mL of Great Salt Lake water using spring water for dilution? Assume the salt concentration from the spring water is negligible.

Do your calculations here.

13. Lead ion can be precipitated out of solution according to the following reaction.

$$Pb^{2+}(aq) + Na_2CrO_4(aq) \rightarrow PbCrO_4(s) + 2\,Na^+(aq)$$

What mass of Na_2CrO_4 should be added to 10.0 L of solution that contains 2.78 g/L of Pb^{2+} ion?

Do your calculations here. Include a Solution map.

14. Colligative properties of solutions depend only on the (1) _____ of solute particles present. Therefore, 1 mole of sugar and 1 mole of alcohol, neither of which is ionic, will depress the freezing point or elevate the (2) _____ of a fixed amount of water by an (3) equal/unequal amount. These properties are usually expressed on the basis of a fixed amount of solvent, which is (4) _____, whether water or some other solvent. Other colligative properties include (5) _____, in addition to freezing point depression and boiling point elevation. A mole of an ionic substance such as NaCl will lower the freezing point of a solution (6) _____ as much as a nonionic material since (7) _____ ions are produced for each mole of NaCl. A mole of $CaCl_2$ will lower the freezing point of water (8) _____ times as much as a mole of sugar or urea. For un-ionized and nonvolatile substances, molecular masses can be determined from either of two colligative properties, namely, (9) _____ and (10) _____.

15. How much ethylene glycol, $C_2H_4(OH)_2$, per kilogram of water is needed to lower the freezing from 0°C to −20°C?

The value for K_f is 1.86°C kg/mol. The equation is:

$$\Delta t_f = (K_f)\left(\frac{\text{grams solute}}{\text{molar mass solute}}\right)\left(\frac{1}{\text{kg solvent}}\right)$$

(1) List the known quantities given in the equation.
(2) Rearrange the given equation to solve for "grams solute".
(3) What else do you need to calculate before solving the problem?
(4) How much ethylene glycol is needed?

Do your calculations here.

16. A sample of an organic compound having a mass of 1.50 g lowered the freezing point of 20.0 g of benzene by 2.75°C. The K_f of benzene is 5.1°C kg/mol. Calculate the molar mass of the compound. The equation is:

$$\Delta t_f = (K_f)\left(\frac{\text{grams solute}}{\text{molar mass solute}}\right)\left(\frac{1}{\text{kg solvent}}\right)$$

First, rearrange the given equation to solve for molar mass of the solute. Then make the necessary substitutions and solve for the molar mass of the unknown solute.

Challenge Problems

17. Cadmium ion, which is a toxic metallic ion, can be precipitated from solution according to the following reaction.

$$Cd^{2+}(aq) + K_3PO_4(aq) \rightarrow Cd_3(PO_4)_2(s) + K^+(aq)$$

Balance the equation and, assuming no side reactions, determine what volume of stock 5.00 M K_3PO_4 solution should be diluted to 250. mL in order to precipitate 4.10 g Cd^{2+} ion that is contained in 600. mL of a waste solution.

Do your calculations here.

18. One gram (1.00 g) of glucose ($C_6H_{12}O_6$) dissolves in 1.10 mL of H_2O. What is the concentration in terms of mass percent and molality? Human blood normally contains approximately 0.090% glucose (mass percent). What is the molality of blood glucose?

Assume 1000 g of blood is equivalent to 1000 g of solvent.

Do your calculations here.

ANSWERS TO QUESTIONS AND SOLUTIONS TO PROBLEMS

1. (1) air, oxygen, nitrogen
    (2) lake and river water, oxygen, water
    (3) iodine solution, iodine, carbon tetrachloride
    (4) bleach, sodium hypochlorite, water

2. (1), (2) sodium, potassium   (3), (4) chloride, iodides   (5), (6) silver, mercury (I)
    (7) phosphates   (8) ammonium   (9) sulfides   (10) ammonium   (11) s   (12) s
    (13) s   (14) s   (15) i   (16) s   (17) s   (18) i   (19) i   (20) i   (21) s
    (22) s   (23) s   (24) s   (25) i   (26) i   (27) i   (28) s   (29) i   (30) s

3. (1) unsaturated   (2) unsaturated   (3) saturated   (4) supersaturated
    (5) unsaturated   (6) unsaturated   (7) unsaturated   (8) supersaturated
    (9) unsaturated   (10) saturated   (11) unsaturated   (12) unsaturated

4. (1) polar   (2) attracted   (3) increase   (4) decrease
    (5) increases   (6) decreases   (7) directly   (8) double

5. (1) $CuSO_4$, KNO_3, KCl, Li_2SO_4, HCl
    (2) Li_2SO_4, $CuSO_4$, KCl, HCl, KNO_3
    (3) 20 g solute in 50 g of water is equivalent to 40 g of solute per 100 g of water. Looking at Figure 14.4, we see that at 10°C, only HCl forms an unsaturated solution (and HCl is a gas so even if the solution were saturated you would not see solid anyway) so the saturated solutions of $CuSO_4$, KNO_3, KCl and Li_2SO_4 solutions will all contain undissolved solid. At 50°C only Li_2SO_4 and $CuSO_4$ will form saturated solutions and contain undissolved solid.

6. Four factors are particle size of a solute, stirring, temperature, solution concentration.

7. (1) 0.5 g NaCl, 524.5 g H_2O

   A 0.1000% solution means that 0.1000% of 525.0 g is salt.

   $$\left(\frac{0.1000}{100}\right)(525.0 \text{ g}) = 0.5250 \text{ g NaCl}$$

   The mass of water is 525.0 g − 0.5250 g = 524.4750 g = 524.5 g H_2O

   (2) 270 mL

   You are asked to calculate a volume contained in a solution of a certain concentration; 18% of the total volume is alcohol.

   $$\left(\frac{18}{100}\right)(1.5 \text{ L}) = 0.27 \text{ L} = 270 \text{ mL}$$

   (3) 15 M H_2SO_4

   For any problem involving molarity and quantities of chemicals needed to prepare for various solutions, it is wise to work from the definitions.

   From the problem, we have 3.8 moles of H_2SO_4 in 250. mL of solution.

   $$M = \left(\frac{3.8 \text{ mol}}{250. \text{ mL}}\right)\left(\frac{1000 \text{ mL}}{L}\right) = 15 \text{ M}$$

   (4) 0.65 M KCl

   $$M = \frac{\text{mol of solute}}{\text{L of solution}} = \frac{2.6 \text{ mol}}{4.0 \text{ L}} = 0.65 \text{ M}$$

   (5) 30.00 g sucrose, 1470 g H_2O

   A 2.000% sucrose solution means that 2.000% of the total solution is sucrose.

   $$\frac{2.000}{100}(1500 \text{ g}) = 30.00 \text{ g sucrose}$$

   If 30.00 g of the total mass is sucrose, then the mass of water is

   $$1500 \text{ g} - 30.00 \text{ g} = 1470 \text{ g } H_2O$$

8. (1) 72.0 g NaOH

   molar mass of NaOH = 22.99 + 16.00 g + 1.008 g = 40.00 g/mol

   We need 600. mL of 3.00 M NaF.

   $$M = \frac{\text{g}}{\text{molar mass} \times \text{L}}$$

Rearranging the equation to solve for g

$$g = M \times \text{molar mass} \times L$$

$$g \text{ of NaF} = (3.00\,M)\left(40.00\,\frac{g}{mol}\right)\left(\frac{600.\,mL \times 1\,L}{1000\,mL}\right)$$

We should use the complete units for M so that our answer has the correct units of grams

$$g \text{ of NaF} = \left(3.00\,\frac{mol}{L}\right)\left(40.00\,\frac{g}{mol}\right)\left(\frac{600.\,mL \times 1\,L}{1000\,mL}\right)$$

$$= 72.0\,g\,\text{NaOH}\,(3\,\text{significant figures})$$

(2) 529 g $(NH_4)_2SO_4$

$$\text{molar mass of}\,(NH_4)_2SO_4 = (2 \times 14.01\,g) + (8 \times 1.008\,g) + 32.07\,g + (4 \times 16.00\,g)$$

$$= 132.15\,\frac{g}{mol}$$

You need 4.00 L of 1.00 M NH_4NO_3

$$g = M \times \text{molar mass} \times L$$

$$= \left(1.00\,\frac{mol}{L}\right)\left(132.15\,\frac{g}{mol}\right)(4.00\,L)$$

$$= 529\,g\,(NH_4)_2SO_4\,(3\,\text{significant figures})$$

(3) 40. g $CaSO_4$

$$\text{molar mass of}\,CaSO_4 = 40.08\,g + 32.07\,g + (4 \times 16.00\,g)$$

$$= 136.15\,\frac{g}{mol}$$

You are asked for 420 mL of 0.70 M $CaSO_4$

$$g = M \times \text{molar mass} \times L$$

$$= \left(0.70\,\frac{mol}{L}\right)\left(136.15\,\frac{g}{mol}\right)\left(\frac{420\,mL \times 1\,L}{1000\,mL}\right)$$

$$= 40.\,g\,CaSO_4\,(2\,\text{significant figures})$$

(4) 46 g $Ca(NO_3)_2$

$$\text{molar mass of}\,Ca(NO_3)_2 = 40.08\,g + (2 \times 14.01\,g) + (6 \times 16.00\,g)$$

$$= 164.10\,\frac{g}{mol}$$

You are asked for 250 mL of 0.94 M $Ca(NO_3)_2$

$$g = M \times \text{molar mass} \times L$$

$$= \left(0.94\,\frac{mol}{L}\right)\left(164.10\,\frac{g}{mol}\right)\left(\frac{250\,mL \times 1\,L}{1000\,mL}\right)$$

$$= 46\,g\,Ca(NO_3)_2\,(2\,\text{significant figures})$$

(5)   2.1 g NaF

$$\text{molar mass of NaF} = 22.99\,\text{g} + 19.00\,\text{g}$$
$$= 41.99\,\frac{\text{g}}{\text{mol}}$$

You are asked for 0.50 L of 0.10 M NaF

$$\text{g} = \text{M} \times \text{molar mass} \times \text{L}$$
$$= \left(0.10\,\frac{\text{mol}}{\text{L}}\right)\left(41.99\,\frac{\text{g}}{\text{mol}}\right)(0.50\,\text{L})$$
$$= 2.1\,\text{g NaF}\,(2\ \text{significant figures})$$

9.   (1)   0.356 L or 356 mL

You are asked to solve the problem for a volume in units of L. With this in mind, write down the complete equation.

$$\text{M} = \frac{\text{g}}{\text{molar mass} \times \text{L}}$$

You now have to rearrange the equation so that "L" is in the numerator and isolated. Multiply both sides of the equation by "L".

$$(\text{L})(\text{M}) = \frac{\text{g}(\text{L})}{\text{molar mass}(\text{L})}$$

Next, cancel the L's of the right side and divide both sides of the equation by M which gives you:

$$\frac{(\text{L})(\text{M})}{(\text{M})} = \frac{\text{g}(\text{L})}{\text{molar mass}(\text{M})(\text{L})}$$
$$(\text{L}) = \frac{\text{g}}{\text{molar mass}(\text{M})}$$

Remember that M is the same as moles per liter and, when you divide by a fraction, the denominator of the fraction comes up to the numerator.

For example: $\dfrac{2}{1/2} = \dfrac{(2)(2)}{1} = 4$

In similar fashion we can write our last equation as

$$(\text{L}) = \frac{\text{g}}{\text{molar mass}\left(\dfrac{\text{moles}}{\text{liter}}\right)} = \frac{\text{g}(\text{L})}{\text{molar mass}(\text{moles})}$$

Determine the molar mass of $K_2Cr_2O_7$, substitute the data in the equation, cancel units and solve for L.

$$\text{molar mass of } K_2Cr_2O_7 = (2 \times 39.10\,\text{g}) + (2 \times 52.00\,\text{g}) + (7 \times 16.00\,\text{g}) = 294.20\,\text{g/mol}$$

Therefore,

$$\text{L} = \frac{130.\,\text{g}(\text{L})}{294.2\,\dfrac{\text{g}}{\text{mol}}(1.24\,\text{mol})} = 0.356\,\text{L or 356 mL}$$

— 113 —

(2)   1.43 L $NaHCO_3$

$$NaHCO_3 = 22.99\text{ g} + 1.008\text{ g} + 12.01\text{ g} + (3 \times 16.00\text{ g}) = 84.01\,\frac{\text{g}}{\text{mol}}$$

$$L = \frac{\text{g}}{\text{molar mass} \times M}$$

This equation is solved just like the one above in 10 (1)

$$= \frac{(48.0\text{ g})(1\text{ L})}{\left(84.01\,\frac{\text{g}}{\text{mol}}\right)(0.400\text{ mol})} = 1.43\text{ L}$$

(3)   0.31 moles of HCl

For this type of problem, go back to the simple first relationship since you are not asked a question involving grams or molar mass. The defining equation for molarity is

$$M = \frac{\text{mol}}{\text{L}}$$

Therefore,

$$\text{moles} = L \times M$$

This is what you have been asked to find in the above problem.

$$\text{moles of HCl} = (2.5\text{ L})\left(0.124\,\frac{\text{mol}}{\text{L}}\right) = 0.31\text{ mol}$$

10.   (1) $K_2Cr_2O_7$ $\text{moles} = \left(\dfrac{250\text{ mL} \times 1\text{ L}}{1000\text{ mL}}\right)\left(0.22\,\dfrac{\text{mol}}{\text{L}}\right) = 0.055\text{ mol}$

(2) Na_2CO_3 $\text{moles} = \left(\dfrac{1500\text{ mL} \times 1\text{ L}}{1000\text{ mL}}\right)\left(1.4\,\dfrac{\text{mol}}{\text{L}}\right) = 2.1\text{ mol}$

(3) $HClO_4$ $\text{moles} = (3.15\text{ L})\left(0.75\,\dfrac{\text{mol}}{\text{L}}\right) = 2.4\text{ mol}$

(4) $CuSO_4 \cdot 5H_2O$ $\text{moles} = \left(\dfrac{857\text{ mL} \times 1\text{ L}}{1000\text{ mL}}\right)\left(0.66\,\dfrac{\text{mol}}{\text{L}}\right) = 0.57\text{ mol}$

11.   In each case, we must solve the relationship for the initial volume (V_1), which is the quantity of the concentrate reagent. Even though we dilute the reagent with water, the number of moles of chemical remains the same; the equation is a simple equality.

(1)   For the first problem, we dilute 15 M NH_4OH (ammonium hydroxide) to 2.5 L of 5.0 M NH_4OH.

Our equation is

$$(V_1)(M_1) = (V_2)(M_2)$$
$$V_1 = \frac{(V_2)(M_2)}{M_1}$$

Substituting in the equation

$$V_1 = \frac{(2.5\text{ L})(5.0\text{ M})}{(15\text{ M})}$$

$V_1 = 0.83$ L or 8.3×10^2 mL of 15 M NH_4OH diluted to 2.5 L with water

(2) H_3PO_4 You have to change 150 mL into L before substituting in the equation.

$$V_1 = \frac{(0.150\,\text{L})(0.30\,\cancel{M})}{(14.6\,\cancel{M})}$$
$$V_1 = 0.0031\,\text{L} = 3.1\,\text{mL}$$

(3) HCl You have to change 500. mL into L first

$$V_1 = \frac{(0.500\,\text{L})(0.25\,\cancel{M})}{(12\,\cancel{M})}$$
$$V_1 = 0.010\,\text{L} = 10.\,\text{mL}$$

(4) H_2SO_4 You have to change 500. mL into L first.

$$V_1 = \frac{(0.250\,\text{L})(1.25\,\cancel{M})}{(18\,\cancel{M})}$$
$$V_1 = 0.017\,\text{L} = 17\,\text{mL}$$

12. 100. L

This is a dilution problem and we can use the equation

$$V_1 M_1 = V_2 M_2$$

where
$V_1 = 500.\,\text{mL} = 0.500\,\text{L}$
$M_1 = 3.42\,\text{M} = 3.42\,\text{mol/L}$
$M_2 = 0.0171\,\text{M} = 0.0171\,\text{mol/L}$
$V_2 = \text{Unknown final volume}$

Substituting into the equation and solving for V_2

$$V_2 = \frac{(0.500\,\text{L})(3.42\,\cancel{\text{mol/L}})}{(0.0171\,\cancel{\text{mol/L}})}$$
$$= 100.\,\text{L}$$

13. 21.7 g Na_2CrO_4

The reaction equation is balanced as written and says that 1 mole of Pb^{2+} ion reacts with 1 mole of CrO_4^{2-} ion. So, we need to calculate how many moles of Pb^{2+} ion we have and from this calculate the mass of Na_2CrO_4 needed.

Solution map: g/L of Pb^{2+} → g of Pb^{2+} → mol Pb^{2+} → mol $Na_2Cr_2O_4$ → mass(g)of $Na_2Cr_2O_4$

$$\text{moles of } Pb^{2+} = \left(\text{mass } Pb^{2+} \text{ per liter/molar mass } Pb^{2+}\right) \times 10.0\,\text{L}$$
$$= \left(\frac{2.78\,\cancel{\text{g/L}}}{207.2\,\cancel{\text{g/mol}}}\right)(10.0\,\cancel{L}) = 0.134\,\text{mol}$$

In order to precipitate 0.134 moles of Pb^{2+} ion, we must add an equal number of moles of Na_2CrO_4 which we can determine as follows:

$$\text{molar mass of } Na_2CrO_4 = 45.98\,\text{g} + 52.00\,\text{g} + 64.00\,\text{g} + 162.0\,\text{g/mol}$$
$$\text{mass } Na_2CrO_4 = (0.134\,\cancel{\text{mol}})(162.0\,\text{g/}\cancel{\text{mol}}) = 21.7\,\text{g}$$

14.

(1) number
(2) boiling point
(3) equal
(4) 1 kg
(5) vapor pressure lowering

(6) about twice
(7) 2 moles
(8) about three
(9) freezing point depression
(10) boiling point elevation

15. 6.67×10^2 g of ethylene glycol

(1) freezing point pure water $= 0°C$; freezing point solution $= -20°C$; $\Delta t_f = 20°C$; $K_f = \dfrac{1.86°C\,kg}{mol}$; kg solvent $= 1$ (looking for mass glycol per kg of water)

(2) $\Delta t_f = (K_f)\left(\dfrac{\text{grams solute}}{\text{MM solute}}\right)\left(\dfrac{1}{\text{kg solvent}}\right) \rightarrow$ grams solute $= \dfrac{(\Delta t_f)(\text{MM})(\text{kg solvent})}{K_f}$

(3) Need to calculate molar mass (MM) of solute, ethylene glycol, $C_2H_4(OH)_2$.

(4) MM of $C_2H_4(OH)_2 = 2 \times 12.01\,g = 24.02\,g$
$$6 \times 1.008\,g = 6.048\,g$$
$$2 \times 16.00\,g = 32.00\,g$$
$$\text{MM} \qquad 62.07\,g$$

grams solute $= \dfrac{(\Delta t_f)(\text{MM})(\text{kg solvent})}{K_f} = (20.0°\cancel{C})\left(\dfrac{62.07\,g}{\cancel{mol}}\right)(1\ \cancel{\text{kg solvent}})\left(\dfrac{\cancel{mol}}{1.86°\ \cancel{C}\ \cancel{kg}}\right)$

grams solute $= 6.67 \times 10^2$ g of ethylene glycol

16. 140 g/mol

We are asked in this problem to determine the molar mass of a compound from freezing point depression data. In order to solve the equation, we need to express the amount of solvent in units of kilograms and then make the necessary substitutions of the other values.

140 g/mol

$$\text{MM solute} = \dfrac{(K_f)(\text{g solute})}{(\Delta t_f)(\text{kg solvent})}$$

Knowns: mass solute $= 1.50$ g; $\Delta t_f = 2.75°C$; mass solvent $= 20.0$ g benzene $K_f = 5.1°C$ kg/mol; MM solute $= ??$ (solve for this)

$$\text{MM solute} = \left(\dfrac{5.1°\ \cancel{C}\ \cancel{kg}}{mol}\right)\dfrac{(1.50\,g)}{(2.75°\ \cancel{C})(0.0200\ \cancel{kg})} = 140\ \text{g/mol}$$

17. Balanced equation is $3\,Cd^{2+}(aq) + 2\,K_3PO_4(aq) \rightarrow Cd_3(PO_4)_2(s) + 6\,K^+(aq)$. Volume of 5.00 M of K_3PO_4 required is 4.86 mL. The equation states that 3 moles of Cd^{2+} require 2 moles of K_3PO_4 to precipitate. The number of moles of Cd^{2+} ion present is equal to

$$\dfrac{4.10\,g\ Cd^{2+}}{112.4\ \text{g/mol}} = 0.0365\ \text{mol}\ Cd^{2+}$$

We will need the following number of moles of K_3PO_4.

$$\text{Moles } K_3PO_4 = \left(0.0365 \text{ mol Cd}^{2+}\right) \frac{2 \text{ mol } K_3PO_4}{3 \text{ mol Cd}^{2+}} = 0.0243 \text{ mol}$$

We next use dilution equation $V_1M_1 = V_2M_2$ where $V_1 = $ unknown volume of stock 5.00 M solution and $M_1 = 5.00$ M.

Since 250. mL of the diluted K_3PO_4 is added to 600. mL of the Cd^{2+} solution, the final volume (V_2) will equal 850. mL (0.850 L).

$$V_2 = 250. \text{ mL} + 600. \text{ mL} = 0.850 \text{ L}$$

$$M_2 = \frac{0.0243 \text{ mol}}{0.850 \text{ L}} = 0.0286 \text{ M}$$

$$V_1 = \frac{(0.850 \text{ L})(0.0286 \text{ M})}{5.00 \text{ M}}$$

$$= 0.00486 \text{ L} = 4.86 \text{ mL}$$

18. Mass percent of glucose solution is 47.6%; molality is 5.05 m. Blood glucose molality is 0.50 m. The molar mass of glucose is 180.2 g/mol. A solution of 1.00 g glucose in 1.10 mL of H_2O is equal to a mass percent concentration of

$$\frac{1.00 \text{ g glucose}}{(1.10 \text{ g}) + (1.00 \text{ g})} \times 100 = 47.6\%$$

We can dissolve $\dfrac{1000.}{1.10}$ g of glucose in 1 kg of water.

This amount is equal to 909 g of glucose or $\dfrac{909 \text{ g}}{180.2 \text{ g/mol}} = 5.04 \text{ mol}$

The molality is therefore 5.04 m.

Human blood has a concentration of 0.090% or 0.90 g per 1000 g of blood. Using the assumption that 1000 g of blood is equivalent to 1000 g of solvent, the molality can be calculated as follows:

$$\frac{0.90 \text{ g}}{180.2 \text{ g/mol}} = 0.0050 \text{ mol}$$

$$\text{molality} = \frac{0.0050 \text{ mol}}{1 \text{ kg solvent}} = 0.0050 \text{ m}$$

Across

2. A mixture without a uniform composition.
6. Yellow solid nonmetallic element (atomic number 16).
7. An element that is ductile and malleable is a _____.
10. One shouldn't do crossword puzzles with a _____.
11. Binary compounds with a metallic element have names ending in _____.
12. Number of protons in hydrogen nucleus.
13. Chemical symbol for tantalum.
15. Chemical symbol for cobalt.
16. Elements whose d and f orbitals are being filled.
17. Formula for nitrogen oxide.
18. Light energy is sometimes considered to exist as a _____.
20. Chemical symbol for element number 95.
22. A pair of electrons forms a chemical _____ between two atoms.
24. Chemical symbol for osmium.
26. A positively charged ion.
27. Chemical symbol for element number 10.

Down

1. The answer to a mathematical problem is often called the _____.
3. Oxygen is an example of an _____, the basic building blocks of matter.
4. Nitrogen exists in what physical state as the uncombined element.
5. Chemical symbol for first element in period number 3.
8. Heat and light are forms of _____.
9. Element number 54.
10. Chemical symbol for element number 84.
13. The element whose chemical symbol is Sn.
14. The smallest indivisible unit of matter.
15. When elements combine chemically, they form a new substance called a _____.
19. A negatively charged ion is called an _____.
21. One Avogadro's Number of a substance dissolved in enough water to give 1 liter of solution is a 1 _____ solution.
23. Chemical symbol for element number 28.
25. Chemical symbol for element number 34.

Answers are found at the end of Chapter 20.

CHAPTER FIFTEEN

Acids, Bases, and Salts

SELECTED CONCEPTS REVISITED

Acids and bases have been defined in three different ways (and by three different sets of scientists whose names are associated with the different definitions). Note that these definitions simply supplement each other. In this course, you will most likely be dealing mainly with the Bronsted-Lowry definition.

Arrhenius acid – releases H^+ into solution
Bronsted-Lowry acid – a proton (H^+) donor
Lewis acid – an electron-pair acceptor
Arrhenius base – releases OH^- into solution
Bronsted-Lowry base – a proton (H^+) acceptor
Lewis base – an electron-pair donor

An acid and its conjugate base are different structurally only by an H^+ (e.g. HCl and Cl^-, or NH_4^+ and NH_3). The same is true for a base and its conjugate acid (e.g. OH^- and H_2O).

Electrolytes are substances capable of producing ions when dissolved in water either by dissociation or ionization. The terms strong and weak describe qualitatively how many ions are produced per molecule or unit of the substance. Acids and bases are also described using the strong and weak terminology. For strong electrolytes, the generic equation $MX \rightarrow M^+ + X^-$ essentially goes to completion. For weak electrolytes, the reaction is written as an equilibrium reaction and the equilibrium lies to the left, for example, $MX \rightleftharpoons M^+ + X^-$.

The higher the H^+ concentration the more acidic the solution and therefore the lower the pH. At a pH of 7, the $[H^+] = [OH^-]$. A pH of less than 7 means that $[H^+] > [OH^-]$ and the solution is acidic.

When writing ionic equations the idea is to show in what form the majority of each species actually exists in solution. For example, since acetic acid (CH_3CO_2H) is a weak acid, when a sample of acetic acid is dissolved in water, most of it is in the molecular form with only a very small percentage in the ionic form. Therefore, acetic acid appears as CH_3CO_2H in a net ionic equation. However, when HCl (g) is dissolved in water, all the HCl ionizes to form H^+ and Cl^- and therefore HCl (aq) would be represented by H^+ and Cl^- in an ionic equation.

$[H^+]$ is the same as $[H_3O^+]$. H^+ is generally associated with ≥ 1 water molecules, so some books/professors use H^+ and H_3O^+ interchangeably.

pH is a logarithmic scale. Therefore a difference of only one pH unit is really a ten-fold difference in acidity.

COMMON PITFALLS TO AVOID

Salt does not necessarily mean "table salt" or NaCl. Many different ionic compounds may be referred to as salts, that is, they can be made from the reaction of an acid with a base.

Pure water is not considered to be an electrolyte and is not considered to be ionic. Puddles, lakes and pools tend to be very good conductors of electricity but that is because of the impurities (of which some are ionic) dissolved in the water.

Do not forget to determine the number of ions in solution per unit or per molecule when dealing with the colligative properties of ionic solutions. Colligative properties depend on the number of particles in solution and each ion counts as a particle.

Not all titrations result in a solution of pH 7. The pH of the solution once the titration is complete depends on the acid and the base used. If a strong acid and a strong base are reacting then the end pH is usually 7, however if a weak acid or base is involved, the end pH is generally not 7.

Please make sure you are familiar with how to use the log and inverse log functions (10^x) on your calculator.

Charges and atoms must be balanced for all properly balanced chemical equations.

Remember that each ionic compound (the ones you will meet) is generally composed of only one type of cation and one type of anion. If there are more than two elements present in the compound, most likely a polyatomic ion is involved.

Use your state symbols to help determine which compounds get broken down into ions for ionic equations. Only compounds followed by (aq) are majority ionic in aqueous solution. Molecules followed by (s), (l), or (g) remain intact and are not written as ions.

When dealing with significant figures for logs, remember that the number in front of the decimal in the pH does not count as a significant figure.

RECAP SECTION

You have likely been exposed to acids and bases already, especially in the lab; in Chapter 15 you will learn in more detail the different definitions, Arrhenius, Bronsted-Lowry, and Lewis, of acids and bases, as well as their general reactions. Given an acid and a base precursor, the salt produced may be predicted and the corresponding chemical equation may be written either in full molecular form, or as a net ionic equation. The properties of electrolytes, their ionization, dissociation, and strength, distinguish them from nonelectrolytes. You need to become familiar with both the actual calculation of the pH of a solution from the hydrogen ion concentration, as well as understand pH and relative acidity. Titration, a fairly old technique used to determine concentration, generally involves neutralization reactions; understand the process, do not simply memorize the mathematical formula. How acid rain forms and its effects on society are also covered in this chapter.

SELF-EVALUATION SECTION

1. There are several different definitions of acids and bases according to various theories. Match the chemist's name with the appropriate phrase:

   (1) Arrhenius   a. Acids are proton donors

   (2) Brønsted-Lowry   b. Acids are electron pair acceptors

   (3) Lewis   c. Acids are hydrogen-containing substances that dissociate to produce H^+

   Designate the following properties as characteristic of an acid or a base.

   (4) changes red litmus to blue _____

   (5) soapy feeling _____   (6) sour taste_____

2. Identify the conjugate acid-base pairs in the following equations.

   (1) $H_2BO_3^- + H_3O^+ \rightarrow H_3BO_3 + H_2O$

   (2) $NH_3 + HBr \rightarrow NH_4^+ + Br^-$

   (3) $HSO_4^- + H_3O^+ \rightarrow H_2SO_4 + H_2O$

   (4) $HC_2H_3O_2 + H_2O \rightarrow H_3O^+ + C_2H_3O_2^-$

   (5) $NH_4^+ + H_2O \rightarrow NH_3 + H_3O^+$

3. Complete and balance the following reactions between HCl and the chemicals given:

(1) HCl and NH_4OH

(2) HCl and Zn metal

(3) HCl and CaO

(4) HCl and Na_2CO_3

4. (1) What is meant by amphoteric?

(2) Complete the equation for the reaction of H_3PO_4 and $Al(OH)_3$, an amphoteric hydroxide.

$$Al(OH)_3 + H_3PO_4 \rightarrow$$

(3) What was $Al(OH)_3$ acting as in the above equation, an acid or a base?

5. Neutralization reactions. Identify the reactants as either *acids* or *bases*.

(1) $H_2SO_4 + Ca(OH)_2 \rightarrow CaSO_4 + 2\ H_2O$
___   ___

(2) $NH_4OH + HClO_3 \rightarrow NH_4ClO_3 + H_2O$
___   ___

Dissociation reactions. Identify the acids and bases in each reaction according to the Brønsted-Lowry theory. Examine both the reactants and the products.

(3) $NH_4^+ + H_2O \rightleftarrows NH_3 + H_3O^+$
___ ___ ___ ___

(4) $C_2H_3O_2^- + H_2O \rightleftarrows HC_2H_3O_2 + OH^-$
___ ___ ___ ___

(5) $Al^{3+} + H_2O \rightleftarrows Al(OH)^{2+} + H^+$
___ ___ ___ ___

6. In which of the following situations does the solution contain an electrolyte (E) or nonelectrolyte (NE)?

(1) The solution contains an acid, such as H_2SO_4. _____

(2) The solution is a nonconductor of electric current. _____

(3) The solution contains only electrically neutral molecules. _____

(4) The solution contains molecules, such as acetic acid, which have reacted with water to produce ions. _____

(5) The solution contains gaseous oxygen. _____

(6) The solution conducts an electric current. _____

(7) The solution contains electrically charged ions. _____

7. Identify the following compounds as strong electrolytes (s) or weak electrolytes (w).

(1) $HC_2H_3O_2$ (6) $HClO_4$

(2) HF   (7) $CaCl_2$

(3) HClO   (8) HCl

(4) NH_4OH (9) H_3BO_3

(5) $Mg(NO_3)_2$ (10) NaOH

8. Calculate the molarity of the ions in each of the salt solutions listed below. Consider each salt to be 100% dissociated.

(1) 0.25 M $MgCl_2$ (4) 0.035 M $Ca(NO_3)_2$

(2) 0.80 M NH_4NO_3 (5) 1.2 M $Al_2(SO_4)_3$

(3) 2.1 M K_3PO_4

9. Identify the following reactions as either dissociation (D) or ionization (I).

(1) $NaCl \xrightarrow{H_2O} Na^+ + Cl^-$ _____

(2) $HC_2H_3O_2 + H_2O \rightarrow H_3O^+ + C_2H_3O_2^-$ _____

(3) $NH_3 + H_2O \rightarrow NH_4^+ + OH^-$ _____

(4) $Ca(OH)_2 \xrightarrow{H_2O} Ca^{2+} + 2\,OH^-$ _____

(5) $HCl + H_2O \rightarrow H_3O^+ + Cl^-$ _____

10. Rank these common solutions from strong acid to strong base. Indicate which solutions are acidic, close to neutral, and basic.

Solution		pH	Rank in order from most acidic to most basic
(1)	0.1 M HCl	1.0	
(2)	Blood	7.4	
(3)	Lemon juice	2.3	
(4)	Drain cleaner	12.1	
(5)	Vinegar	2.8	
(6)	0.1 M $HC_2H_3O_2$	2.9	
(7)	Milk	6.6	
(8)	Ammonium cleaner	8.2	
(9)	Carbonated water	3.0	
(10)	Tomato juice	4.1	

11. Let's do a little bit more with pH. The mathematical expression for pH indicates that we use the hydrogen ion concentration. For example, when $[H^+] = 1 \times 10^{-3}$ mol/L, the pH is 3.

When the $[H^+] = 1 \times 10^{-9}$ mol/L, the pH is 9.

As you can see, we derive the pH value from the exponent of 10 when the number in front is 1. When the number in front of the exponent is between 1 and 10, the pH is between the power of 10 given and the next lower power.

As an example

$[H^+] = 6 \times 10^{-7}$ mol/L   pH is between 7 and 6

$[H^+] = 5.4 \times 10^{-2}$ mol/L   pH is between 2 and 1

(1)   What is the pH of a solution whose $[H^+] = 4 \times 10^{-8}$ mol/L?
       estimated pH = _____; calculated pH _____

(2)   What is the pH of a solution whose $[H^+] = 3.3 \times 10^{-5}$ mol/L?
       estimated pH = _____; calculated pH _____

(3)   What is the pH of a solution whose $[H^+] = 1 \times 10^{-10}$ mol/L?
       pH = _____

(4)   What is the pH of a solution whose $[H^+] = 1 \times 10^{-7}$ mol/L?
       pH = _____

12.   Indicate whether the following equations are *formula* equations (F), *total ionic* equations (T) or *net ionic* equations (N).

(1) $H_2SO_{4(aq)} + MgCO_{3(aq)} \rightarrow MgSO_{4(aq)} + H_2O_{(l)} + CO_{2(g)}$ _____

(2) $2\,H^+_{(aq)} + 2\,Cl^-_{(aq)} + Na_2O_{(s)} \rightarrow 2\,Na^+_{(aq)} + 2\,Cl^-_{(aq)} + H_2O_{(l)}$ _____

(3) $H^+ + OH^- \rightarrow H_2O$ _____

(4) $2\,HCl_{(aq)} + Ca_{(s)} \rightarrow H_{2(g)} + CaCl_{2(aq)}$ _____

(5) $2\,H^+_{(aq)} + SO_4^{2-}{}_{(aq)} + Mg_{(s)} \rightarrow H_{2(g)} + Mg^{2+}_{(aq)} + SO_4^{2-}{}_{(aq)}$ _____

(6) $2\,OH^-_{(aq)} + Mn^{2+}_{(aq)} \rightarrow Mn(OH)_{2(s)}$ _____

(7) $2\,H^+_{(aq)} + CO_3^{2-}{}_{(aq)} \rightarrow H_2O_{(l)} + CO_{2(g)}$ _____

(8) $2\,OH^-_{(aq)} + 2\,Zn_{(s)} \rightarrow 2\,ZnO_{(aq)} + H_{2(g)}$ _____

(9) $Zn(OH)_{2(s)} + 2\,H^+_{(aq)} + 2\,Cl^-_{(aq)} \rightarrow Zn^{2+}_{(aq)} + 2\,Cl^-_{(aq)} + 2\,H_2O_{(l)}$ _____

(10) $3\,NH_4OH_{(aq)} + FeCl_{3(aq)} \rightarrow Fe(OH)_{3(s)} + 3\,NH_4Cl_{(aq)}$ _____

13.   If you are having trouble with writing net ionic equations, this question is intended to walk you through the process. We will be finding the net ionic equation for the last chemical reaction given in the previous question.

Given : $3\,NH_4OH_{(aq)} + FeCl_{3\,(aq)} \rightarrow Fe(OH)_{3\,(s)} + 3\,NH_4Cl_{(aq)}$

First determine which compounds get broken down into ions. Then figure out the formula of the ions involved. Lastly use coefficients to indicate the stoichiometry of each ion. The last column has the info you would use in the total ionic equation. The first set is filled in for you.

Compound	ions involved	stoichiometry of each ion
$3\,NH_4OH_{(aq)}$	NH_4^+, OH^-	$3\,NH_4^+ + 3\,OH^-$
$FeCl_{3\,(aq)}$	_____,_____	_____
$Fe(OH)_{3(s)}$	_____,_____	_____
$3\,NH_4Cl_{(aq)}$	_____,_____	_____

Total ionic equation: _____

Net ionic equation - eliminate all species that are **identical** (formula including charge, and stoichiometry) on both sides of the total ionic equation. _____

14. Write balanced *formula*, *total ionic*, and *net ionic* equations for the following two equations written in words.

(1) Barium chloride solution plus silver nitrate solution react to form insoluble silver chloride plus barium nitrate solution.

Formula equation

Total ionic equation

Net ionic equation

(2) Solid iron plus copper(II) sulfate solution react to form solid copper plus iron(II) sulfate solution.

Formula equation

Total ionic equation

Net ionic equation

15. 35 mL of 0.20 M HCl is required to titrate 50. mL of $Ca(OH)_2$ solution. What is the molarity of the base?

It may help to write a "reverse" solution map-start with what you want to find and map backwards until you have the knowns:

Do your calculations here.

Challenge Problems

16. Identify the conjugate acid-base pairs in the three step ionization of phosphoric acid, H_3PO_4. Write each equation and indicate the acid-base pairs.

Do your calculations here.

17. You are an analyst working on product quality for a vinegar company. Your work this morning involves 10.00 mL samples of various batches of wine vinegar. The data table for the four titrations looks like this:

Sample volume: 10.00 mL

	Batch A	B	C	D
Volume of 0.2500 M NaOH	33.05 mL	34.75 mL	32.80 mL	34.35 mL

(1) If the acidity is due to acetic acid ($HC_2H_3O_2$) what is the average molarity of the vinegar? (2) Given that the density of the vinegar is 1.005 g/mL, what is the average mass percent concentration? (3) If vinegar must be at least 5.000 mass percent concentration of acetic acid, which sample(s) did not meet the standard.

18. How many grams, theoretically, of $BaSO_4$ will be precipitated when 35 mL of 0.15 M H_2SO_4 is added to 18 mL of 0.43 M $Ba(OH)_2$? What is the limiting reactant? If the actual yield is 1.0 g $BaSO_4$, what is the percent yield?

Do your calculation here.

ANSWERS TO QUESTIONS AND SOLUTIONS TO PROBLEMS

1.   (1) c   (2) a   (3) b   (4) base   (5) base   (6) acid

2.   (1) $H_2BO_3^-$ base, H_3BO_3 acid; H_3O^+ acid, H_2O base
      (2) NH_3 base, NH_4^+ acid; HBr acid, Br^- base
      (3) HSO_4^- base, H_2SO_4 acid; H_3O^+ acid, H_2O base
      (4) $HC_2H_3O_2$ acid, $C_2H_3O_2^-$ base; H_2O base, H_3O^+ acid
      (5) NH_4^+ acid, NH_3 base; H_2O base, H_3O^+ acid

3.   (1) $HCl_{(aq)} + NH_4OH_{(aq)} \rightarrow NH_4Cl_{(aq)} + H_2O_{(l)}$
      (2) $2\,HCl_{(aq)} + Zn_{(s)} \rightarrow ZnCl_{2(aq)} + H_{2(g)}$
      (3) $2\,HCl_{(aq)} + CaO_{(s)} \rightarrow CaCl_{2(aq)} + H_2O_{(l)}$
      (4) $2\,HCl_{(aq)} + Na_2CO_{3(aq)} \rightarrow 2\,NaCl_{(aq)} + H_2O + CO_{2(g)}$

4.   (1)   Amphoteric means the species can act as either an acid or a base, depending on the conditions.
      (2) $Al(OH)_{3(s)} + H_3PO_{4(aq)} \rightarrow AlPO_{4(s)} + 3H_2O_{(l)}$
      (3)   base

5.   (1) H_2SO_4 = acid; $Ca(OH)_2$ = base
      (2) NH_4OH = base; $HClO_3$ = acid
      (3) NH_4^+ = acid; H_2O = base; NH_3 = base; H_3O^+ = acid
      (4) $C_2H_3O_2^-$ = base; H_2O = acid; $HC_2H_3O_2$ = base; OH^- = base
      (5) Al^{3+} = acid; H_2O = base; $Al(OH)^{2+}$ = base; H^+ = acid

6.   (1) E   (2) NE   (3) NE   (4) E   (5) NE   (6) E   (7) E

7.   (1) w   (2) w   (3) w   (4) w   (5) s
      (6) s   (7) s   (8) s   (9) w   (10) s

8.   (1)   0.25 M Mg^{2+}, 0.50 M Cl^- (4)   0.035 M Ca^{2+}, 0.070 M NO_3^-
      (2)   0.80 M NH_4^+, 0.80 M NO_3^- (5)   2.4 M Al^{3+}, 3.6 M SO_4^{2-}
      (3)   6.3 M K^+, 2.1 M PO_4^{3-}

9. (1) Dissociation (D). Ions are present in the crystalline salt, and the water molecules act only as the separation agent. Molten salt as well as a solution of salt conducts electricity.

(2) Ionization (I). Water molecules are necessary to form ions in reaction with $HC_2H_3O_2$. Pure $HC_2H_3O_2$ is a very weak electrolyte.

(3) Ionization (I). When ammonia gas dissolves in water, it undergoes reaction with water molecules to form a solution we call ammonium hydroxide (NH_4OH).

(4) Dissociation (D).

(5) Ionization (I). The bond in $HCl_{(g)}$ is predominately covalent.

10. (1)　　(3)　　(5)　　(6)　　(9)　　(10)　　(7)　　(2)　　(8)　　(4)

The first six (in order above) are acids, next two close to neutral, and the last two are basic.

11. (1) Between 8 and 7; 7.4　　(2) Between 5 and 4; 4.5　　(3) 10　　(4) 7

12. (1)　F　　　(2)　T　　　(3)　F　　　(4)　F　　　(5)　T
　　(6)　N　　　(7)　N　　　(8)　N　　　(9)　T　　　(10)　F

13.
Compound	ions involved	stoichiometry of each ion
$3\,NH_4OH_{(aq)}$	NH_4^+, OH^-	$3\,NH_4^+ + 3\,OH^-$
$FeCl_3$	Fe^{3+}, Cl^-	$Fe^{3+} + 3\,Cl^-$
$Fe(OH)_3$	is not dissociated into ions	
$3\,NH_4Cl$	NH_4^+, Cl^-	$3\,NH_4^+ + 3\,Cl^-$

Total ionic equation:

$$3\,NH_4^+ + 3\,OH^- + Fe^{3+} + 3\,Cl^- \rightarrow Fe(OH)_3 + 3\,NH_4^+ + 3\,Cl^-$$

Net ionic equation - eliminate all species that are **identical** (formula including charge, and stoichiometry) on both sides of the total ionic equation.

$$3\,OH^- + Fe^{3+} \rightarrow Fe(OH)_3$$

14. (1) Formula (F)　$BaCl_{2(aq)} + 2\,AgNO_{3(aq)} + 2\,AgCl_{(s)} + Ba(NO_3)_{2(aq)}$
　　　Total ionic (T)　$Ba^{2+}_{(aq)} + 2\,Cl^-_{(aq)} + 2Ag^+_{(aq)} + 2\,NO^-_{3(aq)} \rightarrow 2\,AgCl_{(s)} + Ba^{2+}_{(aq)} + 2\,NO^-_{3(aq)}$
　　　Net ionic (N)　$Ag^+_{(aq)} + Cl^-_{(aq)} \rightarrow AgCl_{(s)}$

(2) Formula (F)　$Fe_{(s)} + CuSO_{4(aq)} \rightarrow Cu_{(s)} + FeSO_{4(aq)}$
　　　Total Ionic (T)　$Fe_{(s)} + Cu^{2+}_{(aq)} + SO^{2-}_{4(aq)} \rightarrow Cu_{(s)} + Fe^{2+}_{(aq)} + SO^{2-}_{4(aq)}$
　　　Net ionic (N)　$Fe_{(s)} + Cu^{2+}_{(aq)} \rightarrow Cu_{(s)} + Fe^{2+}_{(aq)}$

15. 0.070 M* Equation is $2\,HCl_{(aq)} + Ca(OH)_{2(aq)} \rightarrow CaCl_{2(aq)} + 2\,H_2O_{(l)}$. The number of moles of acid used is twice the number of moles of base. Therefore, if we calculate the number of moles of acid used we can use the mole-ratio method to determine the number of moles of $Ca(OH)_2$ in the 50. mL volume, and thus the concentration or molarity.

$$\text{number of moles of HCl} = 0.035\ \cancel{L} \times \frac{0.20\ \text{mol}}{\cancel{L}}\,HCl$$
$$= 0.0070\ \text{mol}$$

the number of moles of $Ca(OH)_2$ is

$$0.0070\ \cancel{\text{mol HCl}} \times \frac{1\ \text{mol Ca(OH)}_2}{2\ \cancel{\text{mol HCl}}} = 0.0035\ \text{mol Ca(OH)}_2$$

Therefore, 0.0035 mol of $Ca(OH)_2$ was present in 50. mL of $Ca(OH)_2$ solution.

---

*molarity of base ⟵　　mol base　　⟵ mol acid ⟵　　molarity acid
　　　　　　　　volume of base (known)　　　　　　volume acid (knowns)

Molarity of $Ca(OH)_2$

$$M = \frac{mol}{L} = \frac{0.0035 \text{ mol } Ca(OH)_2}{0.050 \text{ L}} = 0.070 \text{ M } Ca(OH)_2$$

16. (1) $H_3PO_4 + H_2O \rightarrow H_3O^+ + H_2PO_4^-$
    (2) $H_2PO_4^- + H_2O \rightarrow H_3O^+ + HPO_4^{2-}$
    (3) $HPO_4^{2-} + H_2O \rightarrow H_3O^+ + PO_4^{3-}$

In each case H_3O^+ and H_2O are a conjugate acid-base pair. For reaction (1) H_3PO_4 is the acid and $H_2PO_4^-$ is the base. For reaction (2) $H_2PO_4^-$ is the acid and HPO_4^{2-} is the base while in (3) HPO_4^{2-} is the acid and PO_4^{3-} is the base.

17. Knowns: titration equation: $NaOH + HC_2H_3O_2 \rightarrow H_2O + NaC_2H_3O_2$
     molarity × volume = moles *that is*, (mol/L)(L) = mol
     mole ratio of acetic acid to sodium hydroxide is 1:1
     molar mass of $HC_2H_3O_2$ = 60.06 g/mol
     mass percent acetic acid = $\left(\dfrac{\text{mass of acetic acid in vinegar}}{\text{mass of vinegar}}\right)(100)$
     density of vinegar = 1.005 g/ml

(1) Need to find avg molarity of vinegar (i.e., mol acetic acid per L of vinegar solution)

   Plan: volume NaOH and molarity NaOH → moles NaOH → moles acetic acid in sample
           → avg moles acetic acid in all samples → avg molarity of vinegar

   Calculate: moles NaOH used = V (in L) × M = (0.03305 L) $\left(\dfrac{0.2500 \text{ mol}}{1 \text{ } L}\right)$ (for Batch A)

	Batch A	Batch B	Batch C	Batch D
mol NaOH used:	0.008263	0.008688	0.008200	0.008588
mol $HC_2H_3O_2$ (in 10.00 ml):	0.008263	0.008688	0.008200	0.008588

   average mol $HC_2H_3O_2$ in 10.00 mL = 0.008435 mol

   average molarity = $\dfrac{\text{mol acetic acid}}{\text{L soln}} = \dfrac{0.008435 \text{ mol}}{0.01000 \text{ L}} = 0.8435 \text{ M}$

(2) Need to find mass percent acetic acid in vinegar

   Plan: avg molarity of vinegar → mass of acetic acid in 1 L of vinegar
          density of vinegar → mass of vinegar in 1 L of vinegar → mass percent acetic acid in vinegar

   Calculate: $\dfrac{0.8435 \text{ mol}}{L} \times \dfrac{60.06 \text{ g}}{mol} = 50.66 \text{ g/L} = \text{mass } HC_2H_3O_2 \text{ in 1 L of vinegar}$

   mass of 1 L of vinegar = 1000 $mL \times \dfrac{1.005 \text{ g}}{mL} = 1.005 \text{ g}$

   mass percent = $\dfrac{50.66 \text{ g}}{1005 \text{ g}} \times 100 = 5.041 \% \text{ } HC_2H_3O_2$

(3) Need to find moles of acetic acid in sample when mass percent is 5.000% and compare to moles of acetic acid in the four samples.

   Plan: mass percent acetic acid in vinegar → mass acetic acid in 1 mL vinegar → mass of acetic acid in 10.00 mL vinegar → moles acetic acid in 10.00 mL vinegar → compare with moles of acetic acid in each (10.00 mL) sample in part (1).

Calculate: $\text{mass } \% = \dfrac{\text{mass acetic acid}}{\text{mass vinegar}} \times 100$

rearrange equation to find the mass of acetic acid in 1 mL :

$$\text{mass acetic acid} = (\text{mass } \%) \dfrac{(\text{mass vinegar})}{100} = \dfrac{5.000}{100} \times \dfrac{1.005 \text{ g}}{1 \text{ mL}} = 0.05025 \text{ g/mL}$$

so, mass acetic acid in 10.00 mL vinegar $= \dfrac{0.05025 \text{ g}}{1 \text{ mL}} \times 10 \text{ mL} = 0.5025 \text{ g}$

thus, moles acetic acid in 10.00 mL vinegar $= 0.5025 \text{ g} \times \dfrac{1 \text{ mol}}{60.06 \text{ g}} = 0.008367 \text{ mol}$

Compare this number to the mol $HC_2H_3O_2$ (in 10.00 ml) from each of Batches A-D. Batches A and C do not meet the standard of having at least 5.000% by mass of $HC_2H_3O_2$.

18.　H_2SO_4 is limiting, theoretical yield is 1.2 g $BaSO_4$, percent yield is 83%.

Net ionic equation is $Ba^{2+} + SO_4^{2-} \rightarrow BaSO_{4(s)}$

Number of moles $Ba^{2+} = \left(0.43 \dfrac{\text{mol}}{L}\right)(0.018 \ L) = 0.0077 \text{ mol}$

Number of moles $SO_4^{2-} = \left(0.15 \dfrac{\text{mol}}{L}\right)(0.035 \ L) = 0.0053 \text{ mol}$

Molar mass $BaSO_4 = 233 \dfrac{\text{g}}{\text{mol}}$

Amount $BaSO_4 = \left(233 \dfrac{\text{g}}{\text{mol}}\right)(0.0053 \text{ mol}) = 1.2 \text{ g}$

Percent yield $= \dfrac{\text{Actual yield}}{\text{Theoretical yield}} \times 100 = \dfrac{1.0 \text{ g}}{1.2 \text{ g}} \times 100 = 83\%$

CHAPTER SIXTEEN

Chemical Equilibrium

SELECTED CONCEPTS REVISITED

A reversible reaction is a reaction that can proceed in either direction. That is, as written the reactants may proceed to products or the products can react to reform the reactants.

A reaction has reached equilibrium when the rate of the forward reaction is equal to the rate of the reverse reaction.

Chemists often talk about the equilibrium position in a qualitative way – it describes the relative concentrations of reactants to products. If the equilibrium position lies to the left, it means there are more reactants than products when the reaction has reached equilibrium. The reaction is sometimes referred to as reactant-favored.

Calculating K quantifies the equilibrium position. The value of K provides a ratio of the concentrations and can indicate whether more reactants or products tend to be present at equilibrium. The larger the value of K, the more the reaction proceeded to the right, the more products are present at equilibrium.

$$K = \frac{[products]}{[reactants]} \qquad \text{so for a reaction } 3A + 2B \rightleftharpoons D + 4C; \qquad K = \frac{[D][C]^4}{[A]^3[B]^2}$$

According to Le Chatelier's Principle, a system (reaction) at equilibrium will shift its equilibrium position in an attempt to relieve an added stress. This means if a reactant is removed, the system will try to make more of that reactant; i.e., the equilibrium position will shift to the left.

> If the pressure is increased (if gases present), system tries to decrease its pressure (by reducing number of moles of gas, since gas moles is directly proportional to pressure).
> If moles are added, system tries to use up those moles.
> If the temperature is increased, system tries to lower the temperature by shifting in the direction that uses energy.
> If volume of the flask is decreased (when gases are present), system tries to decrease the volume of the occupied by the particles (shifts in the direction with fewer moles of gas). This is because when volume of flask decreases, the pressure is increased, so the system tries to relieve that pressure by reducing the moles of gases.

Decreasing the volume of reacting <u>gases</u> is equivalent to increasing their concentrations.

The equilibrium constants of certain types of equilibrium reactions have been given special names. Some of these are listed below. You should realize that they are all simply a ratio of the products to the reactants.

K_w = ion product constant for water (the equilibrium constant for the autoionization of water).

K_a = acid dissociation constant (the equilibrium constant for the dissociation or ionization of an acid in water).

K_{sp} = the equilibrium constant for a slightly soluble salt.

The effect caused by the addition of an ion already in a solution at equilibrium is called the common ion effect (the equilibrium position is shifted in the direction that uses the ion as a reactant). Due to the common ion effect, a buffered solution resists changes in pH when small amounts of either acid or base are added to the solution.

COMMON PITFALLS TO AVOID

A reaction at equilibrium has not stopped. Both the forward and the reverse reaction are actually taking place but at the same rate so that no overall change over time is noticed. That is, products are being made as fast as they are being used up to make reactants, so there is no overall change in concentration.

A reaction at equilibrium does not necessarily have equal amounts of reactants to products. In fact the equilibrium position for most reactions is not in the middle. Equilibrium means the rate of both forwards and reverse reactions are equal. The equilibrium position, quantified by the equilibrium constant, K, gives information as to the relative concentrations of reactants to products.

The addition or removal of a catalyst does NOT affect the equilibrium position. A catalyst simply affects how fast the reaction reaches equilibrium but does not affect the equilibrium constant.

Although the system shifts its equilibrium position to relieve an added stress, recognize that it does not fully remove the stress. So if the temperature of a reaction at equilibrium is lowered, the system will shift in the direction that releases energy in an attempt to raise the temperature of the system. Once equilibrium is reestablished, although the temperature of the system will not be back to its original temperature (it will be lower): it would have been even lower without the shift in the equilibrium position.

Stressing the system does NOT change the equilibrium constant, K, unless a temperature change is involved. The actual quantities of each reactant and product changes, but the overall relationship, K, does not.

RECAP SECTION

Chapter 16 discusses two of the most interesting topics in chemistry – chemical equilibrium and chemical kinetics. Kinetics looks at factors that affect the rate of a chemical reaction, such as temperature or surface area, while equilibrium is established when there is no net change in the relative amounts of reactants and products for a reversible reaction. Predicting the changes that occur when concentration, temperature, or volume is changed in a system at equilibrium using Le Chatelier's Principle has important practical value in the chemical industry. You should be comfortable with the meaning and calculation of equilibrium constant. In addition, the equilibrium constant expression may be used to evaluate chemical reactions for ion concentrations and, depending on the situation, the pH of a weak acid or the solubility of a slightly soluble salt. These concepts, along with the function of a buffer solution, are frequently in use by scientists in research and industry.

SELF-EVALUATION SECTION

Most of the examples and problems that we examine for Chapter 16 will have close counterparts in the text material. You should refer to the examples in the text as needed.

1.   Write the following word equations for reversible systems in chemical equation form.

   (1)   Four moles of hydrogen chloride gas plus one mole of oxygen gas react to produce two moles of steam (gaseous water) and two moles of chlorine gas.

   _____

   (2)   One mole of chlorine gas reacts with one mole of liquid water to produce one mole of HC1O solution and one mole of hydrochloric acid solution.

   _____

   (3)   One mole of carbon in the form of coke plus one mole of carbon dioxide gas plus heat react to produce two moles of carbon monoxide gas.

   _____

(4)   One mole of nitrogen gas plus three moles of hydrogen gas react to produce two moles of ammonia gas plus heat.

_____

(5)   Two moles of sulfur dioxide gas plus one mole of oxygen gas react to produce two moles of sulfur trioxide gas.

_____

2.   Using the Principle of Le Chatelier, predict the effect of changing various conditions on the following equilibrium systems.

In each row, the condition that changes has been filled in. Complete the rest of each row with how the system will react to the changing condition.

(1) $2 SO_2(g) + O_2(g) \rightleftharpoons 2 SO_3(g)$

	Equilibrium shift.: (Left, Right, or no)	conc/amt of SO_2	conc/amt of O_2	conc/amt of SO_3	other
a.		SO_2 is added			XXXXXXXX
b.					pressure is increased
c.					catalyst is added
d.					vol of container is decreased

(2) $3 O_2(g) + heat \rightleftharpoons 2 O_3(g)$

	Equilibrium shift.: (Left, Right, or no)	conc/amt of O_2	conc/amt of O_3	other
a.		O_2 is added		XXXXXX
b.			ozone is removed	XXXXXX
c.				reaction is cooled
d.				decrease pressure

(3) $C(s) + CO_2(g) \rightleftharpoons 2 CO(g)$

	Equilibrium shift.: (Left, Right, or no)	conc/amt of C	conc/amt of CO_2	conc/amt of CO	other
a.		C is added			XXXXXX
b.			CO_2 is removed		XXXXXX
c.				CO is removed	XXXXXX
d.					pressure is increased

3. Write the equilibrium constant expression for the following reactions.

Write your answers here.

(1) $PbSO_{4(s)} \rightleftarrows Pb^{2+}_{(aq)} + SO^{2-}_{4(aq)}$

(2) $Ag^+ + 2NH_3 \rightleftarrows Ag(NH_3)^+_{2(aq)}$

(3) $3O_{2(g)} \overset{\Delta}{\rightleftarrows} 2O_{3(g)}$

(4) $C_{(s)} + CO_{2(g)} \overset{\Delta}{\rightleftarrows} 2CO_{(g)}$

(5) $CH_{4(g)} + 2O_{2(g)} \rightleftarrows CO_{2(g)} + 2H_2O_{(g)}$

(6) $HNO_2 \rightleftarrows H^+ + NO_2^-$

(7) $Al^{3+} + OH^- \rightleftarrows Al(OH)^{2+}_{(aq)}$

(8) $2SO_{2(g)} + O_{2(g)} \rightleftarrows 2SO_{3(g)}$

(9) $4NH_{3(g)} + 5O_{2(g)} \rightleftarrows 4NO_{(g)} + 6H_2O_{(g)}$

4. What is the value of K_{eq} for the reaction shown if a 1.0 L flask originally contains 0.80 moles of $A_3B_{(g)}$ at 421°C and, at equilibrium, 50 percent of the A_3B gas has decomposed?

$$A_3B_{(g)} \rightleftharpoons AB_{(g)} + 2A_{(g)}$$

5. (1) What is the concentration of hydrogen ions, H^+, in a 0.25 M solution of nitrous acid, HNO_2? $K_a = 4.5 \times 10^{-4}$

Do your calculations here.

(2) What is the concentration of hydrogen ions, H^+, in a 0.15 M solution of butyric acid, $CH_3(CH_2)_2COOH$? $K_a = 1.52 \times 10^{-5}$

Do your calculations here.

6. (1) What is the H^+ ion concentration and the OH^- ion concentration in a solution of pH 6?

Do your calculations here.

(2) Let's review in a short drill the relationships between pH, pOH, H^+, and OH^- concentrations. The pH of a solution is 8.

What is the pOH? _____
What is the H^+ concentration? _____
What is the OH^- concentration? _____
Is the solution acidic or basic? _____

(3) The pOH of a solution is 10.

What is the OH^- concentration? _____
What is the pH? _____
What is the H^+ concentration? _____
Is the solution acidic or basic? _____

(4) The H^+ concentration is 1×10^{-3} mol/L.
What is the pH? _____
What is the pOH? _____
What is the OH^- concentration? _____
Is the solution acidic or basic? _____

7. (1) What is the pH of a 1×10^{-5} M KOH solution? Assume that the KOH is 100% ionized.

Do your calculations here.

(2) What is the H^+ ion concentration and the OH^- ion concentration in a 0.01 M HBr solution? Assume the HBr is 100% ionized. $K_w = 1 \times 10^{-14}$

Do your calculations here.

8. (1) What is the pH of 94.0 mL of a 0.50 M solution of HC1?

(2) Explain briefly why the solution pH would change after adding 18.0 mL of a 0.0221 M HBr solution to the HC1 solution? Assume volumes are additive, there is no reaction between HBr and HC1, and both are fully ionized.

(3) Calculate the pH of the resulting solution.

9. (1) Which of the following acids would be better used to make a buffer solution? H_2SO_4 or $HC_3H_5O_2$

(2) Why?

(3) What would be an ideal second component of the buffer?

(4) Write the equilibrium equation between the two components of the buffer solution.

(5) With which ion (or molecule) in the equation just written would added H^+ react with? Would added OH^- react with?

10. Look at the compounds listed below. Match up the pairs that would be used for various buffer solutions.

$$HC_2H_3O_2 \quad Na_2HPO_4 \quad H_2CO_3 \quad H_2PO_4^- \quad NaC_2H_3O_2 \quad NaHCO_3$$

11. (1) Lead carbonate, $PbCO_3$, is a slightly soluble salt that dissociates according to the following reaction:

$$PbCO_{3(s)} \rightleftarrows Pb^{2+}_{(aq)} + CO_{3(aq)}^{2-} \quad K_{sp} = 3.3 \times 10^{-14}$$

What is the concentration of CO_3^{2-} in a saturated lead carbonate solution?

Do your calculations here.

(2) Strontium sulfate, $SrSO_4$, is a slightly soluble salt that dissociates according to the following reaction:

$$SrSO_{4(s)} \rightleftarrows Sr^{2+}_{(aq)} + SO_{4(aq)}^{2-} \qquad K_{sp} = 3.8 \times 10^{-7}$$

What is the concentration of Sr^{2+} in a saturated $SrSO_4$ solution?

Do your calculations here.

12. (1) Calcium fluoride, CaF_2, has a solubility of 1.6×10^{-3} g per 100 mL of H_2O. Calculate the K_{sp} remembering that CaF_2 produces two F^- ions and one Ca^{2+} ion when it dissociates. The dissociation equation is

$$CaF_{2(s)} \rightleftarrows Ca^{2+}_{(aq)} + 2\,F^-_{(aq)}$$

Do your calculations here.

(2)   Tin (II) carbonate, $SnCO_3$, has a solubility of 5.73×10^{-3} g/L. Write the K_{sp} expression and then calculate K_{sp}.

Do your calculations here.

(3)   One form of zinc sulfide, ZnS (sphalerite), has a solubility of 6.4×10^{-4} g/L. Calculate the K_{sp}.

Do your calculations here.

13.   Examine the formulas for the compounds shown. Make two lists, one ranking the acids from strongest to weakest and the other ranking the salts from most soluble to least soluble. Then name all of the compounds.

HClO	$K_a = 3.5 \times 10^{-8}$
CaF_2	$K_{sp} = 3.9 \times 10^{-11}$
HCN	$K_a = 4.0 \times 10^{-10}$
$HC_2H_3O_2$	$K_a = 1.8 \times 10^{-5}$
$PbSO_4$	$K_{sp} = 1.3 \times 10^{-8}$
$BaSO_4$	$K_{sp} = 1.5 \times 10^{-9}$

14.   Calculate the H^+ ion concentration and the pH in a 0.10 M H_3PO_4 solution assuming the first ionization reaction is the primary reaction.

$$H_3PO_4 \rightleftarrows H^+ + H_2PO_4^- \qquad K_a = 7.5 \times 10^{-3}$$

Do your calculations here.

Challenge Problems

15. At 720 K, the K_{eq} for the following reaction has a value of 50.5.

$$H_{2(g)} + I_{2(g)} \rightleftarrows 2HI_{(g)}$$

At the given temperature, what is the value for K_{eq} for the reverse reaction.

$$2HI_{(g)} \rightleftarrows H_{2(g)} + I_{2(g)}$$

Calculate the equilibrium concentrations of all 3 species given a 1 L container and 0.0300 moles of HI gas at 720 K.

Do your calculations here.

16. Calculate the pH of a buffer solution prepared by adding 0.10 mole of sodium acetate to 500 mL of a solution labeled 0.40 M acetic acid. K_a is 1.8×10^{-5}.

Do your calculations here.

17. By using net ionic hydrolysis reactions, show why solutions of the following salts are either acidic or basic.

(1) NH_4NO_3 – acidic

(2) K_2CO_3 – basic

(3) NaCN – basic

(4) $Al_2(SO_4)_3$ – acidic

18. Table 16.5 in the text shows that when 0.010 mol HCl or 0.010 mol NaOH is added to 1000 mL of water, the, pH changes by 5 pH units. For each of the following 1000 mL mixtures would you expect a similarly large or a smaller pH change when 0.010 mol NaOH is added.

(1) mixture made from equal amounts of 0.020 M $HClO_4$ and 0.020 M $NaClO_4$.

(2) mixture made from equal amounts of 0.020 M KNO_3 and 0.020 M HNO_3.

(3) mixture made from equal amounts of 0.040 M H_2SO_3 and 0.020 M NaOH.

(4) mixture made from equal amounts of 0.010 M HF and 0.010 M HBr.

ANSWERS TO QUESTIONS AND SOLUTIONS TO PROBLEMS

1. (1) $4HCl_{(g)} + O_{2(g)} \rightleftarrows 2H_2O_{(g)} + 2Cl_{2(g)}$

   (2) $Cl_{2(g)} + H_2O_{(l)} \rightleftarrows HClO_{(aq)} + HCl_{(aq)}$

   (3) $C_{(s)} + CO_{2(g)} + heat \rightleftarrows 2CO_{(g)}$

(4) $N_{2(g)} + 3 H_{2(g)} \rightleftarrows 2 NH_{3(g)} + heat$

(5) $2 SO_{2(g)} + O_{2(g)} + 2 SO_{3(g)}$

2. (1) the conditions that caused the stress to the system are Italicized.

	Equilibrium shift.: (Left, Right, or no)	conc/amt of SO_3	conc/amt of O_2	conc/amt of SO_3	other
a.	shifts right	*SO_2 is added*	decreases	increases	XXXXXX
b.	shifts right (fewer molecules)	decreases	decreases	increases	*pressure is increased*
c.	no effect	no effect	no effect	no effect	*catalyst is added*
d.	shift right	decreases	decreases	increases	*vol of container is decreased*

(2)

	Equilibrium shift.: (Left, Right, or no)	conc/amt of O_2	conc/amt of O_3	other
a.	shifts right	*O_2 is added*	increases	XXXXXX
b.	shifts right	decreases	*ozone is removed*	XXXXXX
c.	shifts left	increases	decreases	*reaction is cooled*
d.	shifts left (towards having more molecules)	increases	decreases	*decrease pressure*

(3)

	Equilibrium shift.: (Left, Right, or no)	conc/amt of C	conc/amt of CO_2	conc/amt of CO	other
a.	shifts right	*C is added*	decreases	increases	XXXXXX
b.	shifts left	increases	*CO_2 is removed*	*decreases*	XXXXXX
c.	shifts right	decreases	*decreases*	*CO is removed*	XXXXXX
d.	shifts left	increases	increases	decreases	*pressure is increased*

3. (1) $K_{eq} = \left[Pb^{2+}\right]\left[SO_4^{2-}\right]$

$PbSO_4$ is a solid; therefore, it is left out of the K_{eq} expression.

(2) $K_{eq} = \dfrac{\left[Ag(NH_3)_2^+\right]}{\left[Ag^+\right]\left[NH_3\right]^2}$

(3) $K_{eq} = \dfrac{\left[O_3\right]^2}{\left[O_2\right]^3}$

(4) $K_{eq} = \dfrac{\left[CO\right]^2}{\left[CO_2\right]}$

Carbon is in the solid state; therefore, it does not change concentration. It can be left out of the equilibrium expression.

(5) $K_{eq} = \dfrac{\left[CO_2\right]\left[H_2O\right]^2}{\left[CH_4\right]\left[O_2\right]^2}$

(6) $K_{eq} = \dfrac{\left[H^+\right]\left[NO_2^-\right]}{\left[HNO_2\right]}$

(7) $\quad K_{eq} = \dfrac{\left[Al(OH)^{2+} \right]}{\left[Al^{3+} \right] \left[OH^- \right]}$

(8) $\quad K_{eq} = \dfrac{[SO_3]^2}{[SO_2]^2 [O_2]}$

(9) $\quad K_{eq} = \dfrac{[NO]^4 [H_2O]^6}{[NH_3]^4 [O_2]^5}$

4. $\quad K_{eq} = 0.064$

The expression for the equilibrium constant, K_{eq}, would be:

$$K_{eq} = \dfrac{[A]^2 [AB]}{[A_3B]}$$

If one-half of the A_3B has decomposed at equilibrium, we can write our equilibrium conditions in tabular form.

Substance	A_3B	A	AB
Initial Conditions	0.80 M	0 M	0 M
Final Conditions	0.40 M	0.80 M	0.40 M

The equilibrium concentration of A is twice that of AB according to the reaction. Substituting into the expression, for K_{eq} we have

$$K_{eq} = \dfrac{[A]^2 [AB]}{[A_3B]} = \dfrac{(0.80)^2 (0.40)}{(0.40)} = 0.064$$

5. (1) 1.1×10^{-2} M or mol/L

The formula for nitrous acid is HNO_2, so the first step is to write out the ionization equation. Then formulate the equilibrium expression.

$$HNO_2 \rightleftarrows H^+ + NO_2^-$$

$$K_a = \dfrac{[H^+] [NO_2^-]}{[HNO_2]}$$

$$4.5 \times 10^{-4} = \dfrac{[H^+] [NO_2^-]}{[HNO_2]}$$

Now you are ready to determine your strategy for making substitutions in the equation. One way to keep everything straight is to set up a small table such as you did for gas law problems.

Substance	H^+	NO_2^-	HNO_2
Initial Conditions	0 M	0 M	0.25 M
Final Conditions	X	X	0.25 – X

Each molecule of HNO_2 that ionizes produces 1 H^+ ion and 1 NO_2^- ion. Therefore, let their equilibrium concentration be "X" so that the amount of HNO_2 left at equilibrium will be 0.25 M – X. Substituting in the equation

$$4.5 \times 10^{-4} = \dfrac{[X][X]}{[0.25 - X]}$$

If X is small compared with 0.25 M, it is possible to simply the equation.

$$4.5 \times 10^{-4} = \frac{[X]^2}{0.25}$$

$$1.1 \times 10^{-4} = X^2$$

To find the value for X, take the square root of 1.1×10^{-4}. Refer to a math text or ask your instructor if you have difficulty with square roots.

Therefore,

$$X = 1.1 \times 10^{-2} \text{ M or mol/L} = [H^+]$$

X is small compared to 0.25 so you were justified in making your assumption.

(2) 1.5×10^{-3} M or mol/L

The formula for butyric acid is $CH_3(CH_2)_2COOH$, so the first step is to write out the ionization equation. After that formulate the equilibrium expression. The ionizable hydrogen in this case is shown on the right.

$$CH_3(CH_2)_2COOH \rightleftarrows CH_3(CH_2)_2COO^- + H^+$$

$$K_a = \frac{[CH_3(CH_2)_2COO^-][H^+]}{CH_3(CH_2)_2COOH}$$

Set up a table as before.

Substance	H^+	$CH_3(CH_2)_2COO^-$	$CH_3(CH_2)COOH$
Initial Conditions	0 M	0 M	0.15 M
Final Conditions	X	X	0.15 – X

Substituting in our equilibrium expression, you have the following equation

$$1.52 \times 10^{-5} = \frac{[X][X]}{[0.15 - X]}$$

Assuming the X is small compared with 0.15, you can simplify your equation.

$$1.52 \times 10^{-5} = \frac{X^2}{0.15}$$

$$2.3 \times 10^{-6} = X^2$$

Taking the square root of both sides without difficulty is possible since the roots are divisible by 2. Therefore,

$$X = 1.5 \times 10^{-3} \text{ M or mol/L}$$

X is small compared to 0.15 so you were justified in making your assumption.

6.   (1) $H^+ = 1 \times 10^{-6}$ M or mol/L, $OH^- = 1 \times 10^{-8}$ M or mol/L. We can determine the H^+ ion concentration directly from the pH and then use the ion product constant of water to calculate the OH^- ion concentration.

The pH value is a whole number, which means that the H^+ concentration is 1 times a negative amount of 10. The value of the exponent is the same as the pH value. Therefore, if the pH is 6, the H^+ ion concentration is 1×10^{-6} M or mol/L. Substituting this value into the ion product constant expression

$$[H^+][OH^-] = K_w = 1 \times 10^{-14}$$

$$[1 \times 10^{-6}][OH^-] = 1 \times 10^{-14}$$

$$[OH^-] = \frac{1 \times 10^{-14}}{1 \times 10^{-6}} = 1 \times 10^{-8} \text{ M or mol/L}$$

(2)   pOH $= 6$

    $[H^+] = 1 \times 10^{-8}$ M

   $[OH^-] = 1 \times 10^{-6}$ M

Basic

(3) $[OH^-] = 1 \times 10^{-10}$ M

      pH $= 4$

    $[H^+] = 1 \times 10^{-4}$ M

Acidic

(4)   pH $= 3$

    pOH $= 11$

   $[OH^-] = 1 \times 10^{-11}$ M

Acidic

7.   (1)   pH $= 9$

There are at least two ways to solve this problem, By one method, we first need to find out what the H^+ concentration is. We can use the ion product constant expression to do this.

$$[H^+][OH^-] = K_w = 1 \times 10^{-14}$$

$$[H^+][1 \times 10^{-14}] = 1 \times 10^{-14}$$

$$[H^+] = \frac{1 \times 10^{-14}}{1 \times 10^{-5}} = 1 \times 10^{-9} \text{ M or mol/L}$$

We can easily convert this value to pH since the value in front of the exponential value is 1 (one). Therefore, the pH is 9. Another way to solve the problem is to use the relationship

$$\text{pH} + \text{pOH} = 14$$

Converting OH^- concentration into pOH, we have

$$1 \times 10^{-5} = \text{pOH of 5}$$

Then, we can substitute the pOH value

$$\text{pH} + 5 = 14$$
$$\text{pH} = 14 - 5 = 9$$

(2) $H^+ = 1 \times 10^{-2}$ M or mol/L, $OH^- = 1 \times 10^{-12}$ M or mol/L.

The problem states that the HBr is completely ionized. The equation would be

$$HBr \rightarrow H^+ + Br^-$$

This means that the H^+ concentration is the same as the HBr solution concentration or 0.01 M. In scientific notation, 0.01 M. would be 1×10^{-2} M. To determine the OH^- concentration, we need to use the ion product constant of water relationship, which states that

$$[H^+][OH^-] = K_w = 1 \times 10^{-14}$$

We know what the concentration of H^+ ion is, so we can make the following substitution into the ion product constant equation

$$[1 \times 10^{-2}][OH^-] = 1 \times 10^{-14}$$

Therefore,

$$[OH^-] = \frac{1 \times 10^{-14}}{1 \times 10^{-2}} = 1 \times 10^{-12} \text{ M or mol/L}$$

Remember, when we are dividing exponents we subtract the denominator exponent from the numerator exponent.

$$-14 - (-2) = -14 + 2 = -12$$

8. (1) 1.699; $pH = -\log[H^+]$; HCl fully ionizes, so $[H^+] = 0.0200$ M; $pH = 1.699$

(2) HBr will fully ionize thus adding more H^+ to the solution making it more acidic, but the volume will increase, effectively diluting the solutions, making them less acidic. It would be difficult to predict the overall effect, other than that the pH will change.

(3) Knowns: 94.0 mL of 0.0200 M HCl

18.0 mL of 0.00744 M HBr

Plan/Map: (working backwards to figure out what is needed. . . .)

$pH \leftarrow [H^+] \leftarrow$ total moles H^+, total volume solution

Calculations:

$$\text{total volume of solution} = 94.0 \text{ mL} + 18.0 \text{ mL} = 112.0 \text{ mL} = 0.1120 \text{ L}$$
$$\text{moles } H^+ \text{ from HCl} = \left(\frac{0.0200 \text{ mol}}{1 \text{ L}}\right)(94.0 \text{ mL})\left(\frac{1 \text{ L}}{1000 \text{ mL}}\right) = 0.00188 \text{ mol}$$
$$\text{moles } H^+ \text{ from HBr} = \left(\frac{0.00744 \text{ mol}}{1 \text{ L}}\right)(18.0 \text{ mL})\left(\frac{1 \text{ L}}{1000 \text{ mL}}\right) = 0.000134 \text{ mol}$$
$$\text{total moles } H^+ = 0.00188 \text{ mol} + 0.000134 \text{ mol} = 0.00201 \text{ mol}$$
$$\text{new}[H^+] = \frac{\text{moles } H^+}{\text{L}} = \frac{0.00201 \text{ mol}}{0.1120 \text{ L}} = 0.0179 \text{ M}$$
$$pH = 1.747 \text{ (less acidic than original solution)}$$

9. (1) $HC_3H_5O_2$

(2) The best buffers are made from weak acids or weak bases. $HC_3H_5O_2$ is a weak acid; sulfuric acid is a strong acid.

(3) A soluble salt of the weak acid would be ideal. For example, $NaC_3H_5O_2$.

(4) $HC_3H_5O_2 \rightleftarrows C_3H_5O_2^- + H_3O^+$

(5) $C_3H_5O_2^-$ would react with added H^+. $HC_3H_5O_2$ would react with any added OH^-.

10. $HC_2H_3O_2$ and $NaC_2H_3O_2$, $\qquad$ Na_2HPO_4 and $H_2PO_4^-$, $\qquad$ H_2CO_3 and $NaHCO_3$

11. (1) $CO_3^{2-} = 1.8 \times 10^{-7}$ M or mol/L

First, write the equilibrium expression with the appropriate value for the solubility product

$$PbCO_3 \rightleftarrows Pb^{2+} + CO_3^{2-}$$

$$K_{sp} = 3.3 \times 10^{-14} = [Pb^{2+}][CO_3^{2-}]$$

Since $PbCO_3$ is a solid whose concentration is unchanging in the equilibrium system you can cancel the term $[PbCO_{3(s)}]$ out of the equation. Since the concentration of Pb^{2+} is equal to the concentration of CO_3^{2-} at equilbrium, you can substitute "X" into the K_{sp} equation for both ion concentrations.

$$3.3 \times 10^{-14} = [X][X]$$
$$3.3 \times 10^{-14} = X^2$$
$$1.8 \times 10^{-7} = X$$

Therefore,

$$\left[Pb^{2+}\right] = \left[CO_3^{2-}\right] = 1.8 \times 10^{-7} \text{ M}$$

(2) $Sr^{2+} = 6.2 \times 10^{-4}$ M

The equation is

$$K_{sp} = 3.8 \times 10^{-7} = \left[Sr^{2+}\right]\left[SO_4^{2-}\right]$$

Since

$$\left[Sr^{2+}\right] = \left[SO_4^{2-}\right] = X$$

Therefore:

$$3.8 \times 10^{-7} = [X][X]$$
$$3.8 \times 10^{-7} = X^2$$
$$6.2 \times 10^{-4} = X$$

Therefore:

$$\left[Sr^{2+}\right] = \left[SO_4^{2-}\right] = 6.2 \times 10^{-4} \text{ M}$$

12. (1) $K_{sp} = 3.2 \times 10^{-11}$

First calculate the molar mass of CaF_2, using a list of atomic masses.

$$\text{molar mass of } CaF_2 = 40.08 \text{ g} + (2 \times 19.00 \text{ g}) = 78.08 \frac{g}{mol}$$

A solubility of 1.6×10^{-3} g per 100 mL of H_2O is the same as 1.6×10^{-2} g per L. Using this number, we can determine the molarity of a saturated solution of CaF_2, assuming the volume of water to be identical with the volume of the solution.

$$\text{Molanty} = \frac{moles}{L}$$

Substituting in the above equation.

$$M = \frac{1.6 \times 10^{-2} \text{ g/L}}{78.08 \text{ g/mol}} = 2.0 \times 10^{-4} \text{ M}$$

$$CaF_{2(s)} \rightleftarrows Ca^{2+}_{(aq)} + 2\,F^-_{(aq)}$$

$$K_{sp} = [Ca^{2+}][F^-]^2$$
$$= [2.0 \times 10^{-4}]\left[2(2.0 \times 10^{-4})\right]^2$$

Remember, there are two F^- ions for each Ca^{2+} ion.

$$K_{sp} = [2.0 \times 10^{-4}]\left[4.0 \times 10^{-4}\right]^2$$
$$= 3.2 \times 10^{-12}$$
$$= 3.2 \times 10^{-11}$$

(2) $K_{sp} = 1.03 \times 10^{-9}$

$$K_{sp} = [Sn^{2+}][CO_3{}^{2-}]$$

First calculate the molar mass of $SnCO_3$, using a list of atomic masses,

$$\text{molar mass of } SnCO_3 = (118.7 \text{ g}) + (12.01 \text{ g}) + (3 \times 16.00 \text{ g}) = 178.7\,\frac{g}{mol}$$

Next, determine the molarity of the saturated solution of $SnCO_3$, then substitute the values in the solubility product equation.

$$\text{Molarity} = \frac{\text{g/L}}{\text{g/mol}} = \frac{\text{mol}}{\text{L}}$$

Substituting in the above equation

$$M = \frac{5.73 \times 10^{-3} \text{ g/L}}{178.7 \text{ g/mol}} = 3.21 \times 10^{-5} \text{ M}$$

$$K_{sp} = [Sn^{2+}][CO_3^{2-}],\, [Sn^{2+}] = [CO_3^{2-}] = 3.21 \times 10^{-5} \text{ M}$$
$$= [3.21 \times 10^{-5}][3.21 \times 10^{-5}]$$
$$= 10.3 \times 10^{-10}$$

$$K_{sp} = 1.03 \times 10^{-9}$$

(3) $K_{sp} = 4.4 \times 10^{-11}$

$$\text{molar mass of } ZnS = 97.44\,\frac{g}{mol}$$

$$M = \frac{\text{g/L}}{\text{g/mol}} = \frac{6.4 \times 10^{-4} \text{ g/L}}{97.44 \text{ g/mol}} = 6.6 \times 10^{-6} \text{ M}$$

$$K_{sp} = [Zn^{2+}][S^{2-}],\, [Zn^{2+}] = [S^{2-}] = 6.6 \times 10^{-6} \text{ M}$$
$$= [6.6 \times 10^{-6}][6.6 \times 10^{-6}]$$
$$= 44 \times 10^{-12}$$

$$K_{sp} = 4.4 \times 10^{-11}$$

13. Acids $HC_2H_3O_2$, acetic acid, is strongest
HClO, hypochlorous acid, is next
HCN, hydrocyanic, is weakest

Salts CaF_2, calcium fluoride, is most soluble
                $PbSO_4$, lead(II) sulfate, is next
                $BaSO_4$, barium sulfate, is least soluble

14. H^+ concentration equals 2.7×10^{-2} M, pH is approximately 1.56. Establish a table as before.

Substance	Initial Conditions	Final Conditions
H_3PO_4	0.10 M	0.10–X
$H_2PO_4^-$	0 M	X
H^+	0 M	X

Substitute the final values into the expression for K_a and assume X has a small value compared to the initial concentration of H_3PO_4.

$$K_a = 7.5 \times 10^{-3} = \frac{[H^+][H_2PO_4^-]}{[H_3PO_4]} = \frac{[X][X]}{[0.10 - X]}$$

$$7.5 \times 10^{-3} = \frac{X^2}{0.10}$$

$$[H^+] = X = \sqrt{7.5 \times 10^{-4}} = 2.7 \times 10^{-2}$$

$$pH = -\log[H^+] = 1.56$$

15. $K_{eq} = 1.98 \times 10^{-2}$, $[H_2] = [I_2] = 4.22 \times 10^{-3}$ M, $[HI] = 2.16 \times 10^{-2}$ M. Quadratic formula yields $[H_2] = [I_2] = 5.88 \times 10^{-3}$ M, $[HI] = 1.83 \times 10^{-2}$ M. The K_{eq} for the reverse reaction is the reciprocal of the K_{eq} for the forward reaction.

$$K_{eq}(\text{reverse}) = \frac{1}{K_{eq}(\text{forward})} = \frac{1}{50.5} = 1.98 \times 10^{-2}$$

The initial concentration of HI is 0.0300 M. Set up a table of values as before

Substance	Initial Conditions	Final Conditions
HI	0.0300 M	0.0300 – X
H_2	0 M	$\dfrac{X}{2}$
I_2	0 M	$\dfrac{X}{2}$

One mole of HI decomposes to produce 0.5 mole each of H_2 and I_2.

$$K_{eq} = \frac{[H_2][I_2]}{[HI]^2} = \frac{\left[\dfrac{X}{2}\right]\left[\dfrac{X}{2}\right]}{[0.0300 - X]^2} = \frac{\left[\dfrac{X^2}{4}\right]}{9.00 \times 10^{-4}} = 1.98 \times 10^{-2}$$

Assume X is small compared to 0.0300 M.

Simplifying

$$\left(1.98 \times 10^{-2}\right)\left(9.00 \times 10^{-4}\right)(4) = X^2$$

$$X^2 = 7.13 \times 10^{-5}$$

$$X = 8.44 \times 10^{-3}$$

If X = 8.44×10^{-3}, then the concentrations of H_2 and I_2 are X/2 or 4.22×10^{-3} M. The concentration of HI is 0.0300 M − 0.00844 M or 0.0216 M.

The value for X is not really small compared to 0.0300 M. Using the quadratic formula, a value for X of 1.17×10^{-2} is obtained rather than 8.44×10^{-3}. In this case the concentrations of H_2 and I_2 would be calculated to be 5.85×10^{3} M and the concentration of HI would be 0.0183 M.

16. pH = 4.4

The equation is $HC_2H_3O_2 \rightleftarrows H^+ + C_2H_3O_2^-$

$$K_a \frac{[H^+][C_2H_3O_2]}{[HC_2H_3O_2]} = 1.8 \times 10^{-5}$$

0.1 mole sodium acetate furnishes 0.1 mole acetate ion.

$$NaC_2H_3O_2 \rightarrow Na^+ + C_2H_3O_2^-$$

$$\frac{0.10 \, mol}{500 \, mL} = \frac{0.20 \, mol}{1000 \, mL} = \frac{0.20 \, mol}{L}$$

Substituting into the equation we have

$$\frac{[H^+][0.20]}{[0.40]} = 1.8 \times 10^{-5}$$

$$H^+ = \frac{(1.8 \times 10^{-5})(0.40)}{0.20} = 3.6 \times 10^{-5} \, M$$

$$pH = -\log[H^+] = 4.4$$

The pH can also be calculated from an equation named the Henderson-Hasselbach equation

$$pH = pKa + \log\frac{[salt]}{[acid]}$$

The answer will be the same either way.

17. (1) Small pH change. This mixture is a buffer system (weak acid and its conjugate base) so a large pH change is not expected when NaOH ts added.

(2) Large pH change. This mixture is made from a strong acid and a neutral salt and so will have the characteristics of a strong acid mixture.

(3) Small pH change. This mixture is a buffer system (weak acid and its conjugate base; the conjugate base was made from the reaction of 1 equivalent of the weak acid with the strong base) so a large pH change is not expected when NaOH is added.

(4) Large pH change. This mixture is made from two strong acids.

WORD SEARCH 2

In the given matrix, find the terms that match the following definitions. Terms may be horizontal, vertical, or on the diagonal. They may also be written forward or backward. Answers are found at the end of chapter 20.

1. A principle that states what happens to an equilibrium system when the conditions are altered.
2. A solution that resists changes in pH.
3. An experimental technique for measuring the volume of one reagent required to react with a measured amount of another reagent.
4. The formation of ions.
5. Solution containing a relatively small amount of solute.
6. A dynamic state where two or more opposing processes are taking place at the same time and at the same rate.
7. A homogeneous mixture of two or more substances.
8. A substance whose aqueous solution conducts electricity.
9. Capable of mixing and forming a solution.
10. The H_3O^+ ion.
11. A solution containing 1 mole of solute per liter of solution is 1.0 _____.
12. A solution containing dissolved solute in equilibrium with undissolved solute.
13. Behaving chemically as either an acid or base.
14. The substance present to the largest extent in a solution.
15. The number of equivalent weights of solute per liter of solution.
16. Incapable of mixing.
17. The substance that is dissolved in a solvent to form a solution.

A	E	G	Y	Y	F	E	H	Y	K	B	D	I	S	P	D	N	J
G	P	I	S	O	L	U	T	E	O	Q	F	Y	D	I	L	K	Z
X	T	D	N	O	A	I	S	Y	C	I	E	V	L	A	H	P	O
C	K	E	B	G	L	O	M	R	I	X	R	U	N	F	O	J	Z
U	M	T	V	A	E	V	Z	M	A	O	T	E	Q	B	Q	Y	M
S	F	A	M	H	C	R	E	D	I	E	R	E	F	F	U	B	Q
D	N	R	U	S	H	E	R	N	C	S	Q	T	U	W	M	J	P
C	O	U	I	J	A	L	B	P	T	L	C	R	C	G	H	Z	W
N	I	T	N	U	T	B	I	N	E	H	A	I	V	E	E	Y	I
X	T	A	O	W	E	J	C	T	O	L	R	T	B	A	L	O	Q
K	A	S	R	D	L	C	V	M	O	E	W	M	Z	L	V	E	N
V	R	G	D	L	I	S	E	M	T	B	L	N	E	C	E	B	L
O	T	F	Y	X	E	I	I	O	N	I	Z	A	T	I	O	N	R
Y	I	G	H	W	R	M	H	Z	A	X	U	F	A	K	T	T	F
I	T	S	V	A	X	P	U	N	O	I	T	U	L	O	S	G	J
J	P	L	Q	Z	M	U	I	R	B	I	L	I	U	Q	E	E	T
M	C	X	S	A	H	T	A	U	D	F	N	Y	P	R	B	R	W

CHAPTER SEVENTEEN

Oxidation-Reduction

SELECTED CONCEPTS REVISITED

Oxidation numbers allow us to keep track of electrons. You will find it much easier to remember the rules for assigning oxidation numbers if you read carefully section 17.1 in your text. This section explains the basis for the assignment of oxidation numbers and how to use electronegativity to help you determine which element to assign first.

Oxidation $\Rightarrow$ loss of electrons $\Rightarrow$ oxidation number increases (gets more positive)
Reduction $\Rightarrow$ gain of electrons $\Rightarrow$ oxidation number decreases (goes left on the number line, more negative)

The oxidizing agent causes oxidation but is itself reduced. That is, the oxidizing agent encourages another substance to lose electrons by taking the electrons away. Since the oxidizing agent gained electrons, it is reduced. The reverse argument applies to reducing agents. Agents are reactants.

When balancing redox reactions, be sure the equation is balanced both atomically and electronically, that is, balance the atoms and the charges. If a redox reaction takes place in aqueous solution, then H_2O is obviously present and can be used as part of a balanced equation. If the reaction is in acidic solution, this means that H^+ ions are also available. In basic solutions, OH^- ions are available to balance the reaction equations.

The higher up in the activity series a species occurs, the more easily it is oxidized. The more easily a species is oxidized, the less likely the product of that oxidation is to be reduced. For example, $K \rightarrow K^+$ is fairly high in the activity series and therefore K is more easily oxidized to K^+ than metals below it in the table (such as $Cu \rightarrow Cu^{2+}$). This means that K^+ is harder to reduce to K than Cu^{2+} is to reduce to Cu.

Your instructor will most likely give you some pneumonics to help remember oxidation-reduction. Here are some of the more popular ones:

OIL RIG   (**O**xidation **I**nvolves **L**oss of electrons, **R**eduction **I**nvolves **G**ain of electrons)
LEO says GER   (**L**oss of **E**lectrons is **O**xidation, **G**ain of **E**lectrons is **R**eduction)

Anode has **ox**idation taking place, while **red**uction occurs at the **cat**hode. (AN OX, RED CAT)

Selected rules for determining oxidation numbers of elements in compounds or ions.
Hydrogen is generally $+1$.
Oxygen is generally -2.
The algebraic sum of the oxidation numbers for all the atoms in a compound is equal to zero.
The algebraic sum of the oxidation numbers for all the atoms in a polyatomic ion is equal to the charge of the ion.

Increasing oxidation number means <u>loss</u> of electrons.

COMMON PITFALLS TO AVOID

Try not to confuse oxidize, reduce, oxidizing agent and reducing agent. The substance being oxidized is losing electrons and is the reducing agent. The species being reduced is gaining electrons and is the oxidizing agent.

Balancing charges does not mean that there must be zero charge on each side of the equation nor that the sum of charges on both sides should equal zero. Balancing charges means that whatever the total charge is on one side of the equation, an equal total charge must be present on the other side of the equation.

RECAP SECTION

The major thrust of Chapter 17 is to give each learner experience at working with chemical equations and manipulating numbers. The ability to assign oxidation numbers to any atom, whether atomic, ionic, or molecular, is key. Redox equations may be balanced by the half-reaction method, or the ion-electron method in acid or base. The activity series of metals may be used to predict reactions or to determine appropriate reagents to use in a galvanic or electrolytic cell. The differences in reactions and functions of those cells are also important, as they have common uses in society. By the end of this chapter, you now have all the skills needed to take a word equation and transform it into a balanced tool to be used for chemical description and calculations, whether ions, molecules, or gases. If you are having difficulty with this material, most questions in this chapter have detail worked answers, so you can follow along and catch where you may have been wrong.

SELF-EVALUATION SECTION

1. Determine the oxidation number for each indicated element in the following compounds or ions:

    (1)   Cr in CrO_3 _____   (2)   Cl in ClO_4^- _____

    (3)   C in CBr_4 _____   (4)   P in P_2O_5 _____

    (5)   N in N_2O _____   (6)   N in NO_2^- _____

    (7)   Mn in MnO_4^- _____   (8)   C in $C_6H_{12}O_6$ _____

    (9)   As in AsO_4^{3-} _____   (10) Br in BrO_3^- _____

    (11) Zn in ZnO_2^{2-} _____   (12) C in $C_2O_4^{2-}$ _____

    (13) C in CH_4 _____   (14) S in H_2SO_3 _____

2. In the following reactions, identify the element that is *reduced*, the element that is *oxidized*, the *reducing agent*, and the *oxidizing agent*.

Oxidation Information

Metallic elements in ionic compounds have a positive charge.

Elements in the free state have an oxidation number of (1) _____.

Hydrogen is generally (2) _____.

Oxygen is generally (3) _____.

Group 1A metals are always (4) _____.

(5) $Cr_2O_{3(s)} + 3 H_{2(g)} \xrightarrow{\Delta} 2 Cr_{(s)} + 3 H_2O_{(l)}$

(6) $2 HNO_{2(aq)} + 2 HI_{(aq)} \rightarrow I_{2(s)} + 2 NO_{(g)} + 2 H_2O_{(l)}$

(7) $2 H_{2(g)} + O_{2(g)} \xrightarrow{\Delta} 2 H_2O_{(l)}$

(8) $5 NaBr + NaBrO_3 + 3 H_2SO_4 \rightarrow 3 Br_2 + 2 Na_2SO_4 + 3 H_2O$

(9) $2 Na_{(s)} + H_2O_{(l)} \rightarrow 2 NaOH_{(aq)} + H_{2(g)}$

3. Consider the following unbalanced reaction; balance it by the change in oxidation number strategy.

$$ClO_{3(aq)}^- + SnO_{2(aq)}^{2-} \rightarrow Cl_{(aq)}^- + SnO_{3(aq)}^{2-}$$

Assign oxidation numbers for each element in each substance in the reaction.
Write and balance the oxidation half-reaction using *only the element being oxidized.*
Write and balance the reduction half-reaction using *only the element being reduced.*
Multiply each half-reaction by an integer such that each uses/produces an equal number of electrons as the other.
Put the coefficients of each half-reaction back into the original unbalanced equation.
Finish balancing, if necessary.

4. Balance the following unionized and ionic equations using the oxidation number strategy. Place correct coefficients in front of each formula.

(1) $H_{(aq)}^+ + Br_{(aq)}^- + SO_{4(aq)}^{2-} \rightarrow Br_{2(l)} + SO_{2(g)} + H_2O_{(l)}$

(2) $NH_{3(g)} + O_{2(g)} \rightarrow NO_{(g)} + H_2O_{(g)}$

(3) $HNO_{2(aq)} + HI_{(aq)} \rightarrow NO_{(g)} + I_{2(s)} + H_2O_{(l)}$

(4) $Mn_{(aq)}^{4+} + Cl_{(aq)}^- \rightarrow Cl_{2(g)} + Mn_{(aq)}^{2+}$

Write your answers here.

5. Consider the following unbalanced reaction; balance it by the ion-reduction strategy.

$$H_2O_{2(aq)} + ClO_{2(aq)} \rightarrow ClO_{2(aq)}^- + O_{2(aq)} \text{ (in basic solution)}$$

Write and balance the oxidation half-reaction using *the entire formula of the ion or molecule.*
Write and balance the reduction half-reaction using *the entire formula of the ion or molecule.*
Balance everything other than O and H.
Add H_2O to balance O, then H^+ to balance H.
If basic, neutralize the H^+ by adding OH^- (equal number must be added to both sides).
Balance charges in each half-reaction by adding electrons.
Multiply each half-reaction by an integer such that each uses/produces an equal number of electrons as the other.
Add the half-reactions together and simplify if necessary.

6. Balance the following redox equations by the ion-electron strategy. (The answer key walks you through the process; however you would not have to write a new equation for every step if solving on a quiz or test-see answer for (3)).

(1) $Zn_{(s)} + NO_{3(aq)}^- \rightarrow Zn_{(aq)}^{2+} + NH_{4(aq)}^+$ (acidic)

(2) $ClO^- + I^- \rightarrow IO_3^- + Cl^-$ (basic)

(3) $MnO_4^- + CN^- \rightarrow MnO_2 + CNO^-$ (basic)

Write your answers here.

7. Using the partial list of the activity series of metals, answer the following questions. For any reaction that proceeds write a balanced equation.

$Mg \rightarrow Mg^{2+} + 2e^-$

$Al \rightarrow Al^{3+} + 3e^-$

$Zn \rightarrow Zn^{2+} + 2e^-$

$Fe \rightarrow Fe^{2+} + 2e^-$

$Ni \rightarrow Ni^{2+} + 2e^-$

$Sn \rightarrow Sn^{2+} + 2e^-$

$Pb \rightarrow Pb^{2+} + 2e^-$

$H_2 \rightarrow 2H^+ + 2e^-$

$Cu \rightarrow Cu^{2+} + 2e^-$

$Ag \rightarrow Ag^+ + e^-$

(1) Will a chemical reaction occur when a $Pb(NO_3)_2$ solution ($Pb^{2+} + 2\ NO_3^-$) is placed in a *copper* container?

(2) Will a chemical reaction occur when a $ZnCl_2$ solution ($Zn^{2+} + 2\ Cl^-$) is placed in a *magnesium* container?

(3) Will a chemical reaction occur when an HCl solution ($H^+ + Cl^-$) is placed in a *tin* can?

(4) Will a chemical reaction occur when an $AgNO_3$ solution ($Ag^+ + NO_3^-$) is placed in a *aluminum* container?

Do your equations here.

8. Aluminum metal is produced almost entirely by the electrolysis of Al_2O_3. The simplified reactions are

$$Al^{3+} + 3e^- \rightarrow Al$$
$$O_{(aq)}^{2-} + C_{(s)} \rightarrow CO_{(g)} + 2e^-$$

In the sketch below, indicate the direction of ion migration and balance the overall equation.

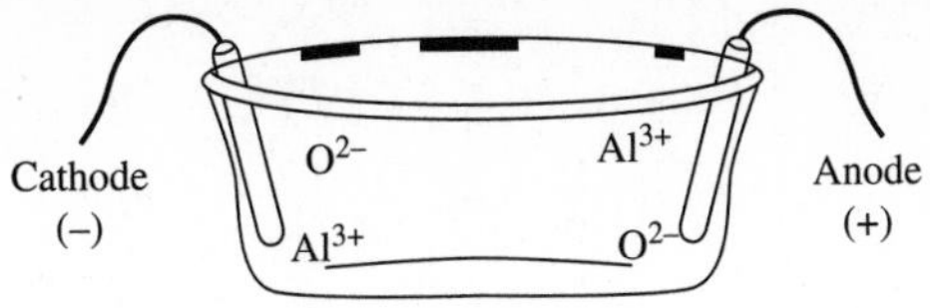

Write your answers here.

9. (1) Write the anode and cathode reactions for the electrolysis of molten $NiBr_2$ with inert electrodes. Then write the balanced overall reaction.

Write your answers here.

(2) Write anode, cathode, and the overall reaction for the electrolysis of molten LiCl with inert electrodes.

Write your answers here.

10. Identify which half-reaction of each pair occurs at the anode and which half-reaction occurs at the cathode.

(1) $Fe^0 \rightarrow Fe^{2+} + 2e^-$

$Ni_2O_3 + 3 H_2O + 2e^- \rightarrow 2 Ni(OH)_2 + 2 OH^-$

(2) $2 H^+ + H_2O_2 + 2e^- \rightarrow 2 H_2O$

$2 I^- \rightarrow I_2 + 2e^-$

Write your answers here.

11. Write the overall reaction for the following pair of half-reactions. Also indicate how many electrons are transferred in the reaction, and which reaction occurs at the anode.

$$Fe^{2+} \rightarrow Fe^{3+} + e^-$$

$$Cr_2O_7{}^{2-} + 14\,H^+ + 6e^- \rightarrow 2\,Cr^{3+} + 7\,H_2O$$

12. Fill in the blank space with a correct term.

The type of cell that uses chemical reactions to produce electrical energy is called a (1) _____ cell. The other type of cell, (2) _____, uses electrical energy to produce a chemical change. A dry cell flashlight battery is an example of a(an) (3) _____.

13. The following reactions are involved in the operation of the lead storage battery. Which is the oxidation reaction and which is the reduction reaction? Give the overall balanced cell reaction.

(1) $Pb \rightarrow Pb^{2+} + 2e^-$

(2) $PbO_2 + 4\,H^+ + 2e^- \rightarrow Pb^{2+} + 2\,H_2O$

14. The super iron battery uses K_2FeO_4 according to the equation
$2\,K_2FeO_4 + 3\,Zn \rightarrow Fe_2O_3 + ZnO + 2\,ZnO_2$
What are the oxidation numbers for each atom in the above equation?

Challenge Problems

15. Balance the following oxidation-reduction equations. The reaction conditions are either acidic or basic, and H^+, OH^- and/or H_2O should be used accordingly to balance the reactions.

(1) $MnO_4^- + VO^{2+} \rightarrow VO_2{}^+ + Mn^{2+}$ (acidic solution)

(2) $P_4 \rightarrow PH_3 + H_2PO_2^-$ (basic solution)

(3) $MnO_2 + SO_3{}^{2-} \rightarrow SO_4{}^{2-} + Mn(OH)_2$ (basic solution)

(4) $Mn_{4(aq)}^- + H_2O_{2(l)} \rightarrow Mn_{(aq)}^{2+} + O_{2(g)}$ (acidic solution)

(5) $Mn^{2+} + HBiO_3 \rightarrow Bi^{3+} + MnO_4^-$ (acidic solution)

(6) $Zn_{(s)} \rightarrow Zn(OH)_{4(aq)}^{2-} + H_{2(g)}$ (basic solution)

16. Determine the oxidation number of the indicated element in each of the following compounds.

(1)	As in $HAsO_3$	_____	(2)	Ti in $Na_2Ti_3O_7$	_____
(3)	Pt in $H_2PtCl_6 \cdot 6H_2O$	_____	(4)	S in $(NH_4)_2S_2O_8$	_____
(5)	S in $Na_2S_2O_3$	_____	(6)	Cr in K_2CrO_4	_____
(7)	Mo in $(NH_4)_2MoO_4$	_____	(8)	Co in $Na_3Co(NO_2)_6$	_____
(9)	B in CaB_4O_7	_____	(10)	C in $H_2C_2O_4$	_____

1. (1) +6 (each O is −2 so the oxidation number of the three O's is −6. CrO_3 is neutral so the oxidation numbers sum to zero. Cr must have an oxidation number of +6.) or $Cr + 3(-2) = 0$, so $Cr = +6$

   (2) +5; sum of oxidation numbers = charge on species; $Cl + 4(-2) = -1$, so $Cl = -1 + 6 = +5$

   (3) +5; $2P + 5(-2) = 0$, so $2P = +10$, so $P = +5$ (4) +4; $C + 4(-1) = 0$, so $C = +4$

   (5) +3; $N + 2(-2) = -1$, so $N = -1 + 4 = +3$ (6) +1; $2(N) + (-2) = 0$, so $2N = +2$, $N = +1$

   (7) 0; $6C + 12(+1) + 6(-2) = 0$, so $C = 0$ (8) +7; $(+1) + Mn + 4(-2) = 0$, so $Mn = +7$

   (9) +5; $As + 4(-2) = -3$, so $As = +5$ (10) +2; $Zn + 2(-2) = -2$, so $Zn = +2$

   (11) +5; $Br + 3(-2) = -1$, so $Br = +5$ (12) +3; $2(C) + 4(-2) = -2$, so $C = +3$

   (13) −4; $C + 4(+1) = 0$, so $C = -4$ (14) +4; $2(+1) + S + 3(-2) = 0$, so $S = +4$

2. (1) 0   (2) +1   (3) −2   (4) +1

   (5) Hydrogen gas is oxidized and is the reducing agent. Chromium is reduced and is the oxidizing agent.

   $$Cr_2O_{3(s)} + 3\,H_{2(g)} \xrightarrow{\Delta} 2\,Cr_{(s)} + 3\,H_2O_{(l)}$$

   First, assign oxidation numbers. $O = -2$ and $H = +1$. Hydrogen gas $= 0$. Chromium metal $= 0$.

   $$\underset{+3\ -2}{Cr_2O_3} + \underset{0}{3\,H_2} \rightarrow \underset{0}{2\,Cr} + \underset{+1\ -2}{3\,H_2O}$$

   What is Cr in Cr_2O_3? If $O = -2$, then each Cr will have to be +3 in order to balance the total amount of negative oxidation value of 3 oxygen. Which element has been oxidized (lost electrons)? Hydrogen has changed from 0 to +1, so it has been oxidized. Cr has changed from +3 to 0, so it has been reduced.

   (6) Iodine is oxidized and is the reducing agent. Nitrogen is reduced and is the oxidizing agent.

   $$\underset{+1\ +3\ -2}{2\,HNO_{2(aq)}} + \underset{+1\ -1}{2\,HI_{(aq)}} \rightarrow \underset{0}{I_{2(s)}} + \underset{+2\ -2}{2\,NO_{(g)}} + \underset{+1\ -2}{2\,H_2O_{(l)}}$$

   Iodine changes from −1 to 0, which is a loss of one electron (oxidation). Nitrogen has gained one electron in changing from +3 to +2.

   (7) Hydrogen gas oxidized and is the reducing agent. Oxygen gas is reduced and is the oxidizing agent.

   $$\underset{0}{2\,H_{2(g)}} + \underset{0}{O_{2(g)}} \xrightarrow{\Delta} \underset{+1\ -2}{2\,H_2O_{(g)}}$$

   Each hydrogen atom has lost an electron, and each oxygen atom has gained two electrons.

   (8) Bromine is both oxidized and reduced.

   $$\underset{+1\ -1}{5\,NaBr} + \underset{+1\ +5\ -2}{NaBrO_3} + \underset{+1\ +6\ -2}{3\,H_2SO_4} \rightarrow \underset{0}{3\,Br_2} + \underset{+1\ +6\ -2}{3\,Na_2SO_4} + \underset{+1\ -2}{3\,H_2O}$$

   The only atoms changing oxidation number are Br. Five Br atoms change from −1 to 0. This is oxidation, a loss of electrons. These Br atoms are the reducing agents. The other Br atom changes from +5 to 0, which is a gain of electrons, or reduction. This Br atom is the oxidizing agent.

   (9) Sodium is oxidized and hydrogen is reduced. Sodium is the reducing agent, and hydrogen is the oxidizing agent.

   $$\underset{0}{2\,Na} + \underset{+1\ -2}{H_2O} \rightarrow \underset{+1\ -2\ +1}{2\,NaOH} + \underset{0}{H_2}$$

   Sodium changes from 0 to +1, an oxidation process. Na is the reducing agent. Hydrogen changes from +1 to 0 for one of the products, hydrogen gas. (Notice that hydrogen is also present in NaOH at the same oxidation state as in H_2O.) Hydrogen is reduced and is the oxidizing agent.

3.　oxidation #'s

$$\underset{+5\ -2}{ClO_{3(aq)}^-} + \underset{+2\ -2}{SnO_{2(aq)}^{2-}} \rightarrow \underset{-1}{Cl_{(aq)}^-} + \underset{+4\ -2}{SnO_{3(aq)}^{2-}}$$

element half-reactions incl e⁻

$$Cl^{5+} + 6e^- \rightarrow Cl^- \qquad\qquad Sn^{2+} \rightarrow Sn^{5+} + 2e^-$$

multiply by whole #'s make # e⁻ equal

$$Cl^{5+} + 6e^- \rightarrow Cl^- \qquad\qquad 3\,Sn^{2+} \rightarrow 3\,Sn^{5+} + 6\,e^-$$

transfer coeff to original

$$ClO_{3(aq)}^- + 3\,SnO_{2(aq)}^{2-} \rightarrow Cl_{(aq)}^- + 3\,SnO_{3(aq)}^{2-}$$

balance other elements
(already balanced)

$$ClO_{3(aq)}^- + 3\,SnO_{2(aq)}^{2-} \rightarrow Cl_{(aq)}^- + 3\,SnO_{3(aq)}^{2-}$$

check

reaction is balanced both electronically (-7 on each side)
and atomically.

4.　(1)　$$4\,H_{(aq)}^+ + 2\,Br_{(aq)}^- + SO_{4(aq)}^{2-} \rightarrow Br_{2(l)} + SO_{2(g)} + 2\,H_2O_{(l)}$$

Assign oxidation numbers and write each half-reaction for oxidation and reduction.

$$\underset{+1}{H_{(aq)}^+} + \underset{-1}{Br_{(aq)}^-} + \underset{+6\ -2}{SO_{4(aq)}^{2-}} \rightarrow \underset{0}{Br_{2(l)}} + \underset{+4\ -2}{SO_{2(g)}} + \underset{+2\ -2}{2\,H_2O_{(l)}}$$

Oxidation　$2\,Br^- \rightarrow Br_2 + 2e^-$
Reduction　$S^{6+} + 2e^- \rightarrow S^{4+}$

Electron gain and loss is balanced. Transfer the coefficient of the half-reactions to the original equation.

$$H_{(aq)}^+ + 2\,Br_{(aq)}^- + SO_{4(aq)}^{2-} \rightarrow Br_{2(l)} + SO_{2(g)} + H_2O_{(l)}$$

Check for balance of atoms. H and O are still out of balance. We need four H^+ on the left side and two H_2O molecules on the right.

$$4\,H_{(aq)}^+ + 2\,Br_{(aq)}^- + SO_{4(aq)}^{2-} \rightarrow Br_{2(l)} + SO_{2(g)} + H_2O_{(l)}$$

Do a check on electrical charges.

$$(4+) + (2-) + (2-) \qquad\qquad 0$$

　left side　　　　　　　　　　right side

(2)　$$4\,NH_{3(g)} + 5\,O_{2(g)} \rightarrow 4\,NO_{(g)} + 6\,H_2O_{(g)}$$

Assign oxidation numbers and write each half-reaction for oxidation and reduction.

$$\underset{-3\ +1}{NH_{3(g)}} + \underset{0}{O_{2(g)}} \rightarrow \underset{+2\ -2}{NO_{(g)}} + \underset{+1\ -2}{H_2O_{(g)}}$$

This equation has a small wrinkle to it. Oxygen is reduced from 0 to -2 and shows up in two molecules on the product side

Oxidation　$N^{3-} \rightarrow N^{2+} + 5e^-$
Reduction　$O_2^{\,0} + 4e^- \rightarrow 2\,O^{2-}$ (from NO and H_2O)

To balance a gain of $4e^-$ and a loss of $5e^-$, we must use 20 as the lowest common multiple

$$4\,N^{3-} \rightarrow 4\,N^{2+} + 20e^-$$
$$5\,O_2 + 20e^- \rightarrow 10\,O^{2-}$$

Put coefficients back in the original equation. Remember that we can only have four NO molecules and that the rest of the oxygen on the product side is in water (need 10 oxygens, so if use 4 for the NO, that leaves a coefficient of 6 for the H_2O).

$$4\,NH_{3(g)} + 5\,O_{2(g)} \rightarrow 4\,NO_{(g)} + 6\,H_2O_{(g)}$$

(3) $2\,HNO_{2(aq)} + 2\,HI_{(aq)} \rightarrow 2\,NO_{(g)} + I_{2(s)} + 2\,H_2O_{(l)}$

Assign oxidation numbers and write each half-reaction for oxidation and reduction.

$$\underset{+1+3-2}{HNO_2} + \underset{+1-1}{HI} \rightarrow \underset{+2-2}{NO} + \underset{0}{I_2} + \underset{+1-2}{H_2O}$$

Oxidation $I^- \rightarrow I_2$

balance $2\,I^- \rightarrow I_2 + 2e^-$ (I^- loses 1 electron)

Reduction $N^{3+} \rightarrow N^{2+}$

balance $N^{3+} + 1e^- \rightarrow N^{2+}$ (N^{3+} gains 1 electron)

Multiply the reduction equation by 2 to balance the electron gain and loss.

$$2\,I^- \rightarrow I_2 + 2e^-$$
$$2\,N^{3+} + 2e^- \rightarrow 2\,N^{2+}$$

Place each half-reaction along with the coefficient back in the original equation.

$$2\,HNO_{2(aq)} + 2\,HI_{(aq)} \rightarrow 2\,NO_{(g)} + I_{2(s)} + 2\,H_2O_{(l)}$$

Balance the remaining H and O atoms.

$$2\,HNO_{2(aq)} + 2\,HI_{(aq)} \rightarrow 2\,NO_{(g)} + I_{2(s)} + 2\,H_2O_{(l)}$$

(4) $Mn^{4+}_{(aq)} + 2\,Cl^-_{(aq)} \rightarrow Cl_{2(g)} + Mn^{2+}_{(aq)}$

This equation should be fairly simple to balance. Assign oxidation numbers.

$$\underset{+4}{Mn^{4+}_{(aq)}} + \underset{-1}{Cl^-_{(aq)}} \rightarrow \underset{0}{Cl_{2(g)}} + \underset{+2}{Mn^{2+}_{(aq)}}$$

Cl^- is oxidized to free elemental chlorine, and Mn^{4+} is reduced to Mn^{2+}.

Oxidation $2\,Cl^-_{(aq)} \rightarrow Cl_{2(g)} + 2e^-$ (1 electron lost per atom, 2 electrons lost per molecule)

Reduction $Mn^{4+}_{(aq)} + 2e^- \rightarrow Mn^{2+}_{(aq)}$ (2 electrons gained per atom)

Put the coefficients in the original equation

$$Mn^{4+}_{(aq)} + 2\,Cl^-_{(aq)} \rightarrow Cl_{2(g)} + Mn^{2+}_{(aq)}$$

The equation is now balanced. Check the electrical charges for the last detail.

$$(4+) + (2-) \quad 2+$$

left side   right side

The equation must have the same electrical charge on both sides as well as the same number of atoms of each element.

5. $H_2O_{2(aq)} + 2\,OH^-_{(aq)} + 2\,ClO_{2(aq)} \rightarrow O_{2(aq)} + 2\,H_2O_{(aq)} + 2\,ClO^-_{2(aq)}$

half-reactions:	$ClO_{2(aq)} \rightarrow ClO_2^-$	$H_2O_{2(aq)} \rightarrow O_{2(aq)}$
Balance all but H, O	$ClO_{2(aq)} \rightarrow ClO_2^-$	$H_2O_{2(aq)} \rightarrow O_{2(aq)}$
(already balanced)		
Balance H, O by adding H^+/H_2O:	$ClO_{2(aq)} \rightarrow ClO_2^-$	$H_2O_{2(aq)} \rightarrow O_{2(aq)} + 2\,H^+$
Balance charges by adding e^-	$e^- + ClO_{2(aq)} \rightarrow ClO_2^-$	$H_2O_{2(aq)} \rightarrow O_{2(aq)} + 2\,H^+ + 2\,e^-$
basic solution	$e^- + ClO_{2(aq)} \rightarrow ClO_2^-$	$H_2O_{2(aq)} + 2\,OH^- \rightarrow O_{2(aq)} + 2\,H_2O + 2\,e^-$
Make #e equal	$2e^- + 2\,ClO_{2(aq)} \rightarrow 2\,ClO_2^-$	
		$H_2O_{2(aq)} + 2\,OH^- \rightarrow O_{2(aq)} + 2\,H_2O + 2\,e^-$
Add	$H_2O_{2(aq)} + 2\,OH^- + 2e^- + 2\,ClO_{2(aq)} \rightarrow O_{2(aq)} + 2\,H_2O + 2\,e^- + 2\,ClO_2^-$	
simplify, check	$H_2O_{2(aq)} + 2\,OH^- + 2\,ClO_{2(aq)} \rightarrow O_{2(aq)} + 2\,H_2O + 2\,ClO_2^-$	

6.   (1) $4\,Zn_{(aq)} + NO_{3(aq)}^- + 10\,H_{(aq)}^+ \rightarrow 4\,Zn_{(aq)}^{2+} + NH_{4(aq)}^+ + 3\,H_2O_{(l)}$

half-reactions: $Zn_{(aq)} \rightarrow Zn_{(aq)}^{2+}$ $NO_{3(aq)}^- \rightarrow NH_{4(aq)}^+$

Balance all but H, O $Zn_{(aq)} \rightarrow Zn_{(aq)}^{2+}$ $NO_{3(aq)}^- \rightarrow NH_{4(aq)}^+$
(already balanced)

Balance H, O by $Zn_{(aq)} \rightarrow Zn_{(aq)}^{2+}$ $NO_{3(aq)}^- + 10\,H_{(aq)}^+ \rightarrow NH_{4(aq)}^+ + 3\,H_2O_{(l)}$
adding H^+/H_2O:

Balance charges by $Zn_{(aq)} \rightarrow Zn_{(aq)}^{2+} + 2\,e^-$ $NO_{3(aq)}^- + 10\,H_{(aq)}^+ + 8e^- \rightarrow NH_{4(aq)}^+ + 3\,H_2O_{(l)}$
adding e^-

Make #e equal $4\,Zn_{(aq)} \rightarrow 4\,Zn_{(aq)}^{2+} + 8\,e^-$ $NO_{3(aq)}^- + 10\,H_{(aq)}^+ + 8e^- \rightarrow NH_{4(aq)}^+ + 3\,H_2O_{(l)}$

Add $4\,Zn_{(aq)} + NO_{3(aq)}^- + 10\,H_{(aq)}^+ + 8e^- \rightarrow 4\,Zn_{(aq)}^{2+} + NH_{4(aq)}^+ + 3\,H_2O_{(1)} + 8e^-$

simplify, check $4\,Zn_{(aq)} + NO_{3(aq)}^- + 10\,H_{(aq)}^+ \rightarrow 4\,Zn_{(aq)}^{2+} + NH_{4(aq)}^+ + 3\,H_2O_{(1)}$

(2) $3\,ClO_{(aq)}^- + I_{(aq)}^- \rightarrow 3\,Cl_{(aq)}^- + IO_{3(aq)}^-$ (in basic solution)

half-reactions: $I^- \rightarrow IO_3^-$ $ClO^- \rightarrow Cl^-$

Balance all but H, O $I^- \rightarrow IO_3^-$ $ClO^- \rightarrow Cl^-$
(already balanced)

Balance H, O by $I^- + 3\,H_2O \rightarrow IO_3^- + 6\,H^+$ $ClO^- + 2\,H^+ \rightarrow Cl^- + H_2O$
adding H^+/H_2O:

Balance charges by $I^- + 3\,H_2O \rightarrow IO_3^- + 6\,H^+ + 6e^-$ $2e^- + ClO^- + 2\,H^+ \rightarrow Cl^- + H_2O$
adding e^-

basic solution so $I^- + 3\,H_2O + 6\,OH^- \rightarrow IO_3^- + 6\,H_2O + 6e^-$ $2e^- + ClO^- + 2\,H_2O \rightarrow$
add OH^- $Cl^- + H_2O + 2\,OH^-$

Make #e equal $I^- + 3\,H_2O + 6\,OH^- \rightarrow IO_3^- + 6\,H_2O + 6e^-$

                    $6e^- + 3\,ClO^- + 6\,H_2O \rightarrow 3\,Cl^- + 3\,H_2O + 6\,OH^-$

Add

$I^- + 3\,H_2O + 6\,OH^- + 6e^- + 3\,ClO^- + 6\,H_2O \rightarrow 3\,Cl^- + 3\,H_2O + 6\,OH^- + IO_3^- + 6\,H_2O + 6e^-$

simplify, check $I^- + 3\,ClO^- \rightarrow 3\,Cl^- + IO_3^-$

(3)   on an exam or quiz, this is usually all you would have to show:

    half reactions: $3e^- + 2\,H_2O + MnO_4^- \rightarrow MnO_2 + 4\,OH^-$ $(\times 2)$
                          $2\,OH^- + CN^- \rightarrow CNO^- + H_2O + 2e^-$ $(\times 3)$

    Overall reaction: $H_2O + 2\,MnO_4^- + 3\,CN^- \rightarrow 2\,MnO_2 + 2\,OH^- + 3\,CNO^-$

7.   (1)   No reaction. Copper is below lead on the series and therefore will not replace lead ions from solution.

    (2)   Reaction. Magnesium is above zinc on the series and will therefore replace zinc ions from solution.

        $Mg + ZnCl_2 \rightarrow Zn + MgCl_2$

    (3)   Reaction. Tin is above hydrogen on the series and will therefore replace hydrogen ions from solution.

        $Sn + 2\,HCl \rightarrow SnCl_2 + H_2$

    (4)   Reaction. Aluminum is above silver on the series and will replace silver ions from solution.

        $Al + 3\,AgNO_3 \rightarrow Al(NO_3)_3 + 3\,Ag$

8. The Al^{3+} ions migrate toward the cathode, and the O^{2-} ions migrate toward the anode. To balance the two half-reactions, balance the electron gain and loss.

$$Al^{3+} + 3e^- \rightarrow Al$$
$$O^{2-} + C \rightarrow CO + 2e^-$$

If you multiply the Al^{3+} half-reaction by 2 and the O^{2-} half-reaction by 3, the electrons will be balanced.

$$2\,Al^{3+} + 6e^- \rightarrow 2\,Al$$
$$3\,O^{2-} + 3\,C \rightarrow 3\,CO + 6e^-$$

Added together: $2\,Al^{3+} + 3\,O^{2-} + 3\,C \rightarrow 2\,Al + 3\,CO$

9. (1) Molten $NiBr_2$ will exist as the following ions $NiBr_2 \xrightarrow{heat} Ni^{2+} + 2\,Br^-$. Oxidation, which is a loss of electrons, occurs at the anode. Br^- is the ion capable of losing electrons. The Ni^{2+} ion needs $2e^-$ (reduction) to become a Ni atom.

Therefore, anode reaction $2\,Br^- \rightarrow Br_2 + 2e^-$

The Ni^{2+} ion needs $2e^-$ (reduction) to become a Ni atom.

$$\text{Cathode reaction} \quad Ni^{2+} + 2e^- \rightarrow Ni^0$$
$$\text{Overall reaction} \quad Ni^{2+} + 2\,Br^- \rightarrow Ni^0 + Br_2$$

(2) LiCl will exist as Li^+ and Cl^- ions in the molten state. Oxidation at the anode will involve the Cl^- ion, but 2 Cl^- ions are needed to balance the equation since Cl_2 is produced.

$$\text{Anode} \quad 2\,Cl^- \rightarrow Cl_2 + 2e^-$$

Reduction at the cathode will use the $2e^-$ available to reduce 2 Li^+ ions to lithium atoms.

$$\text{Cathode} \quad 2e^- + 2\,Li^+ \rightarrow 2\,Li^0$$
$$\text{Overall} \quad 2\,Li^+ + 2\,Cl^- \rightarrow 2\,Li^0 + Cl_2$$

10. (1) Iron goes from the elemental state to 2+. This involves a loss of two electrons and takes place at the anode. Nickel is reduced from the 3+ state to 2+ state at the cathode.

(2) In hydrogen peroxide (H_2O_2), oxygen is in a 1– oxidation state whereas in water oxygen is its usual 2–. This is reduction and occurs at the cathode. I^- loses electrons to reach the elemental state or zero oxidation state, I_2. This is oxidation or the anode reaction.

These examples may be somewhat difficult. Look at them carefully and be sure you are clear on how to determine oxidation states of various atom. Refer to the text if necessary.

11. $6\,Fe^{2+}_{(aq)} + Cr_2O_7^{2-}{}_{(aq)} + 14H^+_{(aq)} \rightarrow 6Fe^{3+}_{(aq)} + 2\,Cr^{3+}_{(aq)} + 7\,H_2O_{(l)}$

Multiply the oxidation half-reaction by 6 to make the number of electrons lost equal to the number of electrons gained according to the reduction half-reaction. Then add the two half-reactions together and simplify.

The oxidation reaction occurs at the anode, and 6 electrons are transferred.

$$6\,Fe^{2+}_{(aq)} \rightarrow 6Fe^{3+}_{(aq)} + 6\,e^-$$

12. (1) voltaic   (2) electrolytic   (3) voltaic

13. Reaction (1) is oxidation, reaction (2) is reduction. Overall reaction:

$$Pb + PbO_2 + 4\,H^+ \rightarrow 2\,Pb^{2+} + 2\,H_2O$$

14.
$$2\,K_2FeO_4 + 3\,Zn \rightarrow Fe_2O_3 + ZnO + 2\,K_2ZnO_2$$
$${}_{+1\ +6\ -2}{}_{0}{}_{+3\ -2}{}_{+2\ -2}{}_{+1\ +2\ -2}$$

K_2FeO_4: each K is $+1$, each O is -2 so $(2)(+1) + x + (4)(-2) = 0$; $x(Fe) = +6$
Zn in its elemental state has an oxidation number of 0
Fe_2O_3: each O is -2 so $2x + 3(-2) = 0$; $x(Fe) = +3$
ZnO: each O is -2 so $x + (-2) = 0$; $x(Zn) = +2$
K_2ZnO_2: each K is $+1$, each O is -2 so $2(+1) + x + 2(-2) = 0$; $x(Zn) = +2$

15. (1) $\quad MnO_4^- + 5\,VO^{2+} + H_2O \rightarrow 5\,VO_2^+ + 2\,H^+ + Mn^{2+}$

(2) $\quad P_4 + 3\,H_2O + 3\,OH^- \rightarrow PH_3 + 3\,H_2PO_2^-$

(3) $\quad MnO_2 + SO_3^{2-} + H_2O \rightarrow Mn(OH)_2 + SO_4^{2-}$

(4) $\quad 5\,H_2O_2 + 2\,MnO_4^- + 6\,H^+ \rightarrow 2\,Mn^{2+} + 5\,O_2 + 8\,H_2O$

(5) $\quad 2\,Mn^{2+} + 5\,HBiO_3 + 9\,H^+ \rightarrow MnO_4^- + 5\,Bi^{3+} + 7\,H_2O$

(6) $\quad Zn_{(s)} + 2\,OH_{(aq)}^- + 2\,H_2O_{(l)} \rightarrow Zn(OH)_{4\,(aq)}^{2-} + H_{2\,(g)}$

16. (1) $+5$ (2) $+4$ (3) $+4$ (4) $+7$ (5) $+2$
    (6) $+6$ (7) $+6$ (8) $+3$ (9) $+3$ (10) $+3$

CHAPTER EIGHTEEN

Nuclear Chemistry

SELECTED CONCEPTS REVISITED

Nuclear reactions involve the nucleus of the atom and therefore the product of a nuclear reaction can be, and often is, a different element. In all the previous reactions you have seen, only the electrons were shifted and the elements involved in the reactant were still part of the product. For a nuclear reaction, the protons, neutrons and electrons may all be involved and therefore the product may involve a different element.

Balancing nuclear reactions involves balancing not the elements but the atomic numbers and mass numbers. Therefore it is important that all reactants and products be written in isotopic notation. The mass numbers are balanced independently of the atomic numbers. The elements or particles are assigned based on the atomic number of the particle. Table 18.1 in your textbook shows the isotopic notation for radioactive and subatomic particles.

There are many terms in this chapter that are similar. Please ensure you understand definitions. For example, nucleons are the particles in the nucleus, a neutron is one of the two types of nuclear particles (the neutral one), and a nuclide refers simply to any isotope of any atom.

Fission involves chain reactions as a neutron is initially captured by a nuclide which then disintegrates, producing two or more neutrons which are then available to be captured to start the process over.

Nuclear fission is the splitting of a nuclide using a neutron.
Nuclear fusion is the joining of two nuclides.
Both fission and fusion release large quantities of energy.

COMMON PITFALLS TO AVOID

A mass of radioactive material does not disappear after two half-lives! The half-life is the time it takes for the radioactive sample to be reduced (decay) to half its current mass. For example, if the half-life of 48 g of a radioactive sample were 1 day, then after 1 day, only 24 g would be left. But the half-life for the 24 g quantity is also one day so at the end of the second day (after a total of two half-lives) 12 g would still be present.

For nuclear equations, do NOT try to balance elements. Add or subtract the appropriate mass and atomic numbers and then use the atomic numbers to determine the element. Also, chemical and nuclear equations are always written as sums, not differences, so you don't subtract particles, but rather, you write the particle as a product instead.

RECAP SECTION

Chapter 18 is a fascinating chapter about a subject that influences us all. We are constantly bombarded by cosmic radiation, nuclear power plants pose problems relating from meltdowns due to natural disasters to disposal of highly reactive wastes, and we may sometimes come into contact with radioactive isotopes during medical treatment. This chapter covers the particles associated with nuclear chemistry, radioactivity, and use half-life to calculate the age of an object. You should be familiar with the three main radioactive emissions of alpha particle, beta particle, and gamma rays and be able to write balanced nuclear equations. There are several instruments used to measure radiation, which can be expressed in a variety of units. The transmutation of elements occurs by both natural and artificial disintegration series, and both nuclear fusion and fission are associated with large energy changes; mass defect and binding energy are both concepts you should be familiar with. Although ionizing radiation on living organism can have significant detrimental effects, radioactivity is not something to fear, but it is something to be discussed with knowledge and respect. Every citizen should know the essential details concerning radioactivity and its current and potential uses.

SELF-EVALUATION SECTION

1. Consider the nuclide $^{80}_{27}X$.

   (1) Fill in the blanks with one of the following choices.
   element symbol, atomic number, mass number

   (a) _____ ⟵ ——————— 80

   $^{80}_{27}X$ ⟶ (c) _____

   (b) _____ ⟵ ——————— 27

   (2) What are the three principal rays or particles coming from the nucleus of a radioactive nuclide?
   (a) _____, (b) _____, and (c) _____.

   (3) What are their symbols (isotopic notation form)

   (4) Assume $^{80}_{27}X$ can decay by all three forms of radioactivity. Write the symbol of the resulting nuclide. Use a different letter to indicate a different element.
   (a) after alpha decay _____
   (b) after beta decay _____
   (c) after gamma radiation _____

2. (1) A 16.0 gram sample of $^{223}_{88}Ra$ takes 11.2 days to decay to 8.0 grams. The 11.2 days is called the _____ of the nuclide.

   (2) How many days would it take the original 16.0 gram sample of $^{223}_{88}Ra$ to decay to 0.5 grams?

3. For each of the following characteristics, place the three common types of radiation in the correct order.

	Most		Least
(1) Ionizing power	_____	_____	_____
(2) Heaviest to lightest	_____	_____	_____
(3) Velocity of radiation	_____	_____	_____
(4) Electrical charge (+)	_____	_____	_____(−)
(5) Penetrating ability of radiation	_____	_____	_____

4. (1) For the following half-life sketch, indicate the mass of the sample after each half-life, and indicate the total time that has elapsed.

   $$64.00\,g \xrightarrow{1045\ years} (a) \rightarrow (b) \rightarrow (c) \rightarrow (d) \rightarrow (e) \rightarrow (f) \rightarrow (g) \rightarrow (h)$$

   (2) After 420 days (six half-lives), the resulting mass of a radioactive nuclide was 0.75 g. What was the original mass and half-life of the material?

5. Fill in the missing atomic number, atomic mass, or missing symbol for the following reactions. You may need to use a periodic table.

   (1) $^{210}_{84}Po \rightarrow ^{206}_{82}Pb$ (5) $^{14}_{7}N + \underline{\quad} \rightarrow ^{14}_{6}C + ^{1}_{1}H$

   (2) $^{209}_{83}Bi + \underline{\quad} \rightarrow ^{210}_{84}Po + ^{1}_{0}n$ (6) $^{14}_{7}N + ^{1}_{0}n \rightarrow ^{12}_{6}C + \underline{\quad}$

(3) $^{35}_{17}Cl + ^1_0n \rightarrow$ _____ $+ ^1_1H$

(7) $^{60}_{28}Ni + ^1_1H \rightarrow ^{57}_{27}Co +$ _____

(4) _____ $+ ^1_0n \rightarrow ^{24}_{11}Na + ^4_2He$

(8) $^7_3Li + ^1_1H \rightarrow$ _____ $+ ^1_0n$

6. Identify the following reactions as nuclear fission or nuclear fusion reactions.

(1) $^2_1H + ^2_1H \rightarrow ^3_2He + ^1_0n$

(2) $^{238}_{92}U + ^1_0n \rightarrow ^{144}_{56}Ba + ^{90}_{36}Kr + 2^1_0n$

(3) $^7_3Li + ^1_1H \rightarrow 2^4_2He$

(4) $^2_1H + ^2_1H \rightarrow ^3_1H + ^1_1H$

(5) $^1_0n + ^{238}_{92}U \rightarrow ^{144}_{54}Xe + ^{90}_{38}Sr + 2^1_0n$

(6) $^3_1H + ^2_1H \rightarrow ^4_2He + ^1_0n$

7. Fill in the blank space or circle the correct response.

High levels of radiation, especially gamma or X rays, are termed (1) _____ radiation. If the dosage is high enough, (2) _____ can occur within several days. The effects of radiation appear to be localized in the (3) _____ of cells. Rapidly growing and dividing cells are (4) most/least susceptible to damage. Long-term or protracted exposure to (5) high/low levels of radiation can lead to health problems at some later time in a person's life. Presently there is concern about the effect that strontium-90, which is chemically similar to the element (6) _____, has on blood cells manufactured in bone marrow. An accumulation of Sr-90 may lead to increased incidence of bone cancer and leukemia, Radiation damage to the nucleus of a cell can affect future generations of a particular species by giving rise to genetic (7) _____. Such events occur when the radiation damages the genetic material – a molecule called (8) _____ – but not severely enough to prevent reproduction.

8. A radioactive nuclide undergoes the following disintegration series:

$\alpha, \beta, \beta, \alpha, \alpha, \alpha, \alpha, \alpha, \gamma, \alpha, \beta, \beta, \alpha, \beta, \beta, \alpha, \alpha, \gamma, \alpha.$

(1) By how much has the nuclide lost/gained in atomic number?

(2) By how much has the nuclide lost/gained in mass number?

9. Write an equation representing

(1) the alpha decay of uranium-238.

(2) the emission of a beta particle that results in bismuth-214

10. Match the names of scientists associated with nuclear chemistry with the appropriate descriptive phrase.

(1) Becquerel   a. Discovered alpha and beta rays

(2) Marie Curie   b. Developed cyclotron

(3) Rutherford   c. Coined word "radioactivity"

(4) E. O. Lawrence   d. Found that uranium salts emit rays

(5) Hahn and Strassmann   e. Reported first instance of nuclear fission

11. Fill in the blank space or circle the correct response.

A magnetic field affects the three principal radioactive rays differently. The beta particle, being (1) positively/negatively charged, will be attracted (2) toward/away from the positive plate. The gamma ray, which has a mass of (3) _____ and an electrical charge of (4) _____, (5) will/will not be attracted by the magnetic field. The alpha ray, which has a charge of (6) _____, will be deflected (7) _____ the negative plate.

12. A piece of a wooden tool found at a Northwest Indian fishing village site has been determined to be approximately 10,000 years old. The ^{14}C in the wood has undergone approximately how many half lives of decay? ^{14}C has a half-life of 5668 years.

Write your answer here.

13. The half life for $^{32}_{15}P$, a common biological radionuclide, is 14.3 days. Starting with 1500 micrograms of $^{32}_{15}P$, how much will you have left after 100 days?

Do your calculations here.

14. Fill in the blank space or circle the correct response.

The mass of an atomic nucleus is (1) more/less than the sum of the masses of the particles that make up the nucleus. The difference in mass is known as the (2) _____. The energy equivalent to this mass (using the equation $E = mc^2$) is called the (3) _____ of the nucleus. This amount of energy would be required to (4) put together/pull apart the particles of a particular nucleus. The higher the binding energy, the (5) more/less stable the nucleus is. In both nuclear fission and fusion reactions, the products have less mass than the reactants. The resultant mass losses are accounted for in the very large quantities of energy that are released.

Challenge Problems

15. Calculate the binding energy for $^{32}_{16}S$ which occurs naturally at an abundance of 95%. The mass of one atom of $^{32}_{16}S$ is known to be 31.97207 amu (atomic mass units). The masses of the elemental particles are as follows:

proton $= 1.007277$ amu　　　　　neutron $= 1.008665$ amu
electron $= 0.0005486$ amu　　　　1 amu $= 1.49 \times 10^{-10}$ J

Do your calculations here.

16. The binding energy of one $^{7}_{3}Li$ atom is 6.258×10^{-12} J. Using the masses of the elemental particles listed in problem 12, calculate the actual mass (in amu) of $^{7}_{3}Li$ atom.

1 amu $= 1.49 \times 10^{-10}$ J

Do your calculations here.

1. (1) (a) mass number   (b) atomic number   (c) element symbol
   (2) (a) alpha   (b) beta   (c) gamma
   (3) alpha, $_2^4$He, beta $_{-1}^0$e, gamma, γ
   (4) (a) $_{25}^{76}$Z (the mass number is 4 less, the atomic number is 2 less and the nuclide therefore has different elemental symbol)
        (b) $_{28}^{80}$Y (the mass number does not change, the atomic number increases by 1 and the nuclide therefore has different elemental symbol)
        (c) $_{27}^{80}$X (gamma radiation has no mass and therefore there is no change in symbol)

2. (1) half-life
   (2) 56 days (5 half-lives: 16.0 — 8.0 — 4.0 — 2.0 —1.0 — 0.5)

3. (1) γ, β, α (2) α, β, γ (3) γ, β, α (4) α, γ, β (5) γ, β, α

4. (1) (a) 160.00 g   (b) 80.00 g   (c) 40.00 g   (d) 20.00 g   (e) 10.00 g   (f) 5.00 g
        (g) 2.50 g   (h) 1.25 g   total elapsed time = 1045 × 8 = 8360 years
   (2) 48.00 g, 70 days   For the mass, do NOT simply take (0.75 × 6)!

   The material was double before each half-life elapsed, so working backwards:
   after 5 half-lives there would have been (0.75 × 2) = 1.50 g
   after 4 half-lives there would have been (1.50 × 2) = 3.00 g
   after 3 half-lives there would have been (3.00 × 2) = 6.00 g
   after 2 half-lives there would have been (6.00 × 2) = 12.00 g
   after 1 half-life there would have been (12.00 × 2) = 24.00 g
   therefore the original sample would have (24.00 × 2) = 48.00 g

5. Missing species would be:

   (1) $_2^4$He   (2) $_1^2$H   (3) $_{16}^{35}$S

   (4) $_{13}^{27}$Al   (5) $_0^1$n   (6) $_1^3$H

   (7) $_2^4$He   (8) $_4^7$Be

6. (1) Nuclear fusion – two light atoms combine to form a heavier nucleus.
   (2) Nuclear fission – heavy, unstable nucleus splits into two smaller fragments under bombardment with neutrons.
   (3) Nuclear fusion   (4) Nuclear fusion   (5) Nuclear fission.

7. (1) acute   (2) death   (3) nucleus   (4) most
   (5) low   (6) calcium   (7) mutations   (8) DNA

8. There are a total of 11 α decays and 6 β decays; γ-radiation will have no effect. 11 alpha decays results in a loss of 44 in mass number and 22 in atomic number. 6 beta decays results in no change in atomic mass and a gain of 6 in atomic number.
   (1) Atomic number will go down by 16.
   (2) Atomic mass will be reduced by 44.

9. (1) $_{92}^{238}$U → $_2^4$He + $_{90}^{234}$Th

   $\left(\text{note : NOT } _{92}^{238}\text{U} - _2^4\text{He} \rightarrow _{90}^{234}\text{Th}\right)$

   (2) $_{82}^{214}$Pb → $_{83}^{214}$Bi + $_{-1}^0$e

10. (1) d   (2) c   (3) a   (4) b   (5) e

11. (1) negatively   (2) toward   (3) none   (4) zero
    (5) will not   (6) 2^+ (7) toward

12. Approximately 2.

The half-life for $^{14}_{6}C$ is 5668 years

13. 12 μg

100 days is 7 half-lives. After this period of time 0.78% of the original material will be left.

0.0078 × 1500 μg = 12 μg

14. (1) less   (2) mass defect   (3) binding energy   (4) pull apart

    (5) more

15. Mass defect is equal to 0.29178 amu which is equal to 4.3475×10^{-11} J/atom of $^{32}_{16}S$.

16. 7.016 amu

The calculated mass of the elemental particles in a $^{7}_{3}Li$ atom gives a value of 7.058137 amu.

The binding energy is equal to 4.2×10^{-2} amu.

CHAPTER NINETEEN

Introduction to Organic Chemistry

SELECTED CONCEPTS REVISITED

Learning the structure of the functional groups is crucial to your knowledge of organic chemistry. The nomenclature, physical properties and chemical properties of organic compounds are all based on the functional groups present in the molecule. Being able to recognize the different functional groups in a molecule therefore gives you a large insight into the potential reactivity of the compound.

Functional groups generally contain polar bonds and/or pi bonds; chemical reactions most often revolve around the functional groups.

Although it may appear as though there are a different set of rules for naming each type of functional group on compounds, the underlying premise and process is actually fairly simple. A worked example given on the next page summarizes the general nomenclature process. Nomenclature requires practice!

A popular question for homework or exams is asking a student to correct an incorrectly named organic compound. Should you be asked to correct the name of a compound, first draw the compound from the name given, then ignore that potentially incorrect name. The most common errors to check for include ensuring that the longest chain is correctly found, that the branches/substituents are numbered from the correct end of the chain, and that the branches/substituents are listed alphabetically in the name. Prefices are not considered when alphabetizing; they do not affect the alphabetical order of the substituents.

Carbon is the backbone of organic chemistry. Covalently bonded, neutral carbon atoms have four bonds. Unless otherwise shown, it is assumed those bonds are to hydrogen. So if given a skeletal structure of C-C-C, the compound is really $CH_3CH_2CH_3$. If given C-C-C-Cl, the compound is really $CH_3CH_2CH_2Cl$.

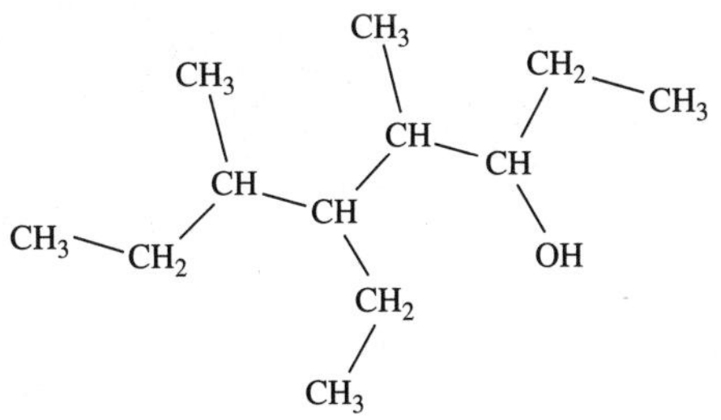

The molecule at left is the same as that at right. The only difference is that the CH_3, CH_2 and CH groups are not shown at right, only their connecting bonds. Therefore, the end of each line and each intersection actually has a carbon atom present.

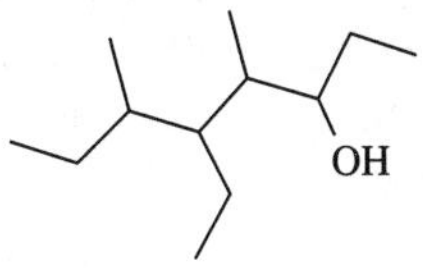

Iso means same . . . structural isomers have same molecular formula, but different structures (atoms connected in different order).

Ethers have two carbon groups attached to the oxygen (carbon chains sandwiching an oxygen). Each carbon chain has to be counted separately. Ethers are named either by simply naming each of the two chains separately as alkyl groups, then adding the word "ether", (i.e., short alkyl chain then long alkyl chain then ether), OR, short chain root name followed by "oxy" then full name of longer chain (alkoxyalkane). E.g., $CH_3CH_2CH_2CH_2OCH_3$ would be methyl butyl ether or methoxybutane (carbon chains of four and one). Carbon chains of two and five, $CH_3CH_2OCH_2CH_2CH_2CH_2CH_3$, would be ethoxypentane or ethyl pentyl ether.

Esters have two carbon chains also, but they are not "equal". One carbon chain has the C=O (originates from the carboxylic acid) and one is just an alkyl group (originates from the alcohol). You need to be able to recognize which side is which in order to name the compound. The regular alkyl group is named first, as an alkyl group. The side with the C=O adds "oate" to the root name. For example, **CH₃CH₂COOCH₂CH₂CH₂CH₃** is named as butyl **propan**oate. Even though the four carbon chain is longer than the chain on the other side, what matters is where the chains are attached. The three carbon chain involves the C=O and so "oate" is added to the "propan" root. The four carbon chain does not have the C=O and so it becomes butyl.

What to look for:	Impact on name:	Example
find the functional group(s) in the molecule such as alkene or ketone groups. E.g., alcohol	determines the ending of the name, such as -ene, -one. E.g., "**ol**".	
find the longest straight chain that includes the functional group. E.g., eight carbons	determines the root of the main part of the name that gets placed before the ending found above. E.g., **octanol**	
number from the end of the straight chain that gives the functional group the lowest possible number (if you have an alkane, number such that the first branch has the lowest possible number). E.g., number from right to left	determines the number that immediately precedes the name you have thus far – the number indicates the position of the FG on the main chain. E.g., **3-octanol**	
find any branches or substituents; number and name them. E.g., branches are on carbons 4,5,6 of the main chain.	E.g., a methyl group on carbon 4 and another on carbon 6 and an ethyl group on carbon 5 will be named as follows: **4-methyl** **6-methyl** **5-ethyl**	
organize the branches in alphabetical order (keeping the numbers assigned to them), then combine any similar branches and prefix the branch name with di, tri, etc. as appropriate.	E.g., 5-ethyl-4-methyl-6-methyl gets shortened to 5-ethyl-**4,6-di**methyl-	
Put the whole name together with the branches first.	E.g., 5-ethyl-4,6-dimethyl-3-octanol	

COMMON PITFALLS TO AVOID

Carbon atoms *cannot* have 5 bonds! Carbon normally has four bonds (satisfies the octet rule). A carbon with only three bonds to it must be charged.

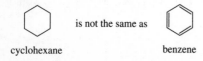

cyclohexane is not the same as benzene

If you see an OH on a molecule, please take a closer look to verify whether it is an alcohol (ROH) or a carboxylic acid (RCOOH).

If you see an OR group on a molecule, please take a closer look to verify whether it is an ether (ROR'), an ester (RCOOR'), or a ketone (RCOR').

Aldehydes are abbreviated as RCHO not RCOH.

Ketones are abbreviated as RCOR' not ROR'. The C bearing the O cannot have any H's attached., (i.e., RCH$_2$OR' represents an ether, not a ketone. RCH$_2$COR' represents a ketone.)

Do not assume that the longest straight chain is written in a straight line. For example, the longest straight chain in the following compound is shown in bold and is a 9-carbon chain.

$$\text{CH}_3\text{CH}_2 - \textbf{CH} - \textbf{CH}_2\textbf{CH}_2\textbf{CH}_2\textbf{CH}_2\textbf{CH}_3$$
$$\textbf{CH}_2\textbf{CH}_2\textbf{CH}_3$$

RECAP SECTION

Chapter 19 has given you a brief look at organic chemistry; the chemistry of carbon. Carbon is limited to having four bonds but is able to form long chains, the most basic of which are the alkanes. Identification of organic compounds by their functional groups, being able to write their structural formulas, and naming them are introduced in this chapter. In addition, common reactions, especially for the more basic functional groups are discussed. The different types of hydrocarbons, ethers, carbonyl compounds, alcohols, and ethers, are among the functional groups discussed. The chemical composition of polymers, the basis for plastics, is also covered. In spite of the tremendous variety of organic compounds, scientists have kept the field manageable by organizing them systematically by functional group. The reaction mechanisms, molecular structure, and intricate synthesis schemes are often similar for members within a functional group. As you continue in science, you will find yourselves regularly confronted with organic chemicals.

SELF-EVALUATION SECTION

1. Fill in the blank space or circle the appropriate response.

    Carbon, with four outer shell electrons, is able to form (1) one/four single (2) ionic/covalent bonds by sharing its electrons with other elements. The bond angles are not planar but describe a (3) cubic/tetrahedral shape. The remarkable ability of carbon to bond to (4) itself/only metals leads to the possibility of long chain compounds, ring compounds, and structures containing single, double, and even triple bonds. A single bond consists of (5) _____ electrons, a double bond of (6) _____ electrons, and a triple bond of (7) _____ electrons. For example, in the formula C$_2$H$_2$, each carbon atom is joined to two hydrogen atoms by a (8) _____ covalent bond, and the two carbon atoms are joined together by a (9) _____ bond.

2. Draw the Lewis structure of carbon tetrabromide, acetylene, and methane. Keep in mind the possibility of multiple bonds.

    Do your structures here.

3. Draw the Lewis structures for 2-iodobutane and pentane.

    Do your structures here.

4. Match the structural formulas and the names for the isomers of heptane. Look for the longest continuous chain of carbon atoms and then the location of the alkyl groups along the chain.

Formulas	Names

(1) $CH_3-CH_2-CH_2-CH_2-CH_2-CH_2CH_3$ 3-methylhexane

(2)
$$CH_3-CH_2-CH_2-CH-CH_3$$
$$|$$
$$CH_2CH_3$$

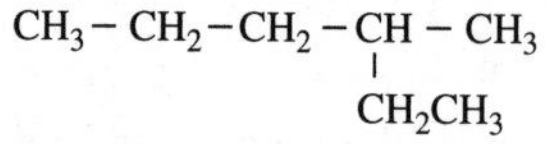

n-heptane

(3)
$$CH_3-CH_2-CH-CH_2-CH_3$$
$$|$$
$$CH_2-CH_3$$

3-ethylpentane

(4)
$$CH_3$$
$$|$$
$$CH_3-C-CH-CH_3$$
$$|\ \ \ \ |$$
$$CH_3\ \ CH_3$$

3,3-dimethylpentane

(5)
$$CH_3$$
$$|$$
$$CH_3-C-CH_2-CH_3$$
$$|$$
$$CH_2-CH_3$$

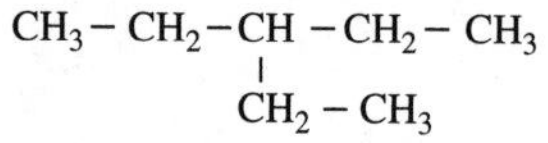

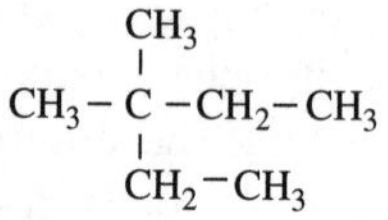

2,2,3-trimethylbutane

5. Name the following alkanes and alkyl halides. Look for the longest carbon chain and name them so that the substituent numbering is as small as possible.

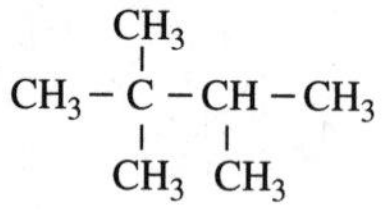

(1) 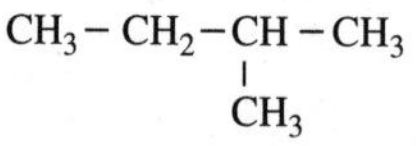
$$CH_3-CH_2-CH-CH_3$$
$$|$$
$$CH_3$$

_____

(2)
$$CH_3\ \ CH_2-CH_2Cl$$
$$|\ \ \ \ |$$
$$CH_3-CH_2-C-CH-CH_3$$
$$|$$
$$CH_3$$

_____

(3)
$$Cl$$
$$|$$
$$CH_3-CH_2-CH-CH_3$$

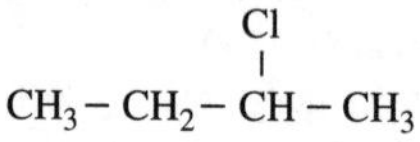

_____

(4) 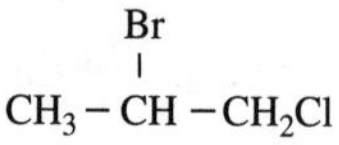
$$Br$$
$$|$$
$$CH_3-CH-CH_2Cl$$

_____

(5) 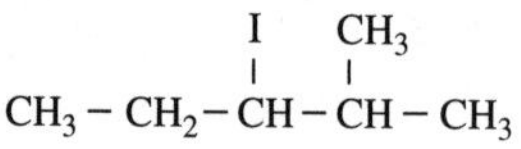
$$I\ \ \ \ CH_3$$
$$|\ \ \ \ |$$
$$CH_3-CH_2-CH-CH-CH_3$$

_____

(6) 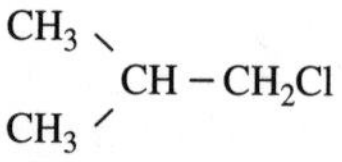
$$CH_3 \searrow$$
$$CH-CH_2Cl$$
$$CH_3 \nearrow$$

_____

(7) 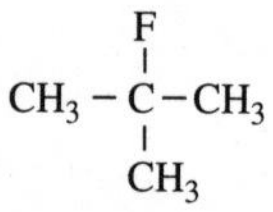
$$F$$
$$|$$
$$CH_3-C-CH_3$$
$$|$$
$$CH_3$$

_____

(8) 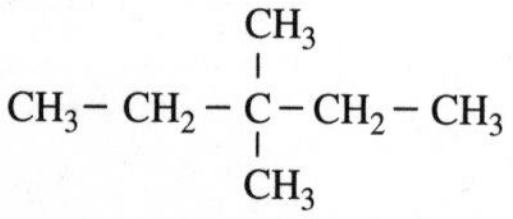
$$CH_3$$
$$|$$
$$CH_3-CH_2-C-CH_2-CH_3$$
$$|$$
$$CH_3$$

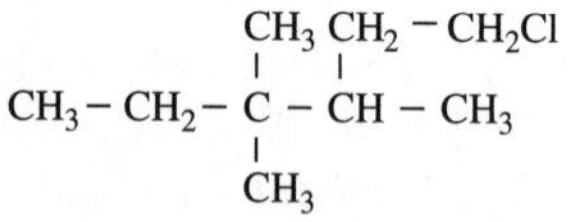

_____

6. Draw the possible structures for the monohydroxy (—OH) isomers of n-heptane. Be sure to eliminate any duplicate structures.

Write your structures here.

7. Fill in the missing members of the alkane, alkyne and aldehyde homologous series. Give the formulas or names or both. Note that there are two formula columns for the aldehydes. Aldehydes are more commonly written as RCHO where the R is an alkyl group. Note the trend of how the "$C_\#H_\#$" before the CHO changes.

Alkanes		Alkynes			Aldehydes		
methane	CH_4	Must have $\geq$ 2C's		_____	HCHO	CH_2O
ethane	C_2H_6	ethyne	C_2H_2	_____	CH_3CHO	C_2H_4O
_____	C_3H_8	propyne	C_3H_4	propanal	C_2H_5CHO	C_3H_6O
butane	_____	_____	_____	butanal	_____	_____
pentane	_____	_____	C_5H_8	pentanal	_____	_____
_____	C_6H_{14}	hexyne	_____	_____	$C_5H_{11}CHO$	_____
_____	_____	_____	C_7H_{12}	_____	$C_6H_{13}CHO$	_____
octane	C_8H_{18}	_____	_____	octanal	_____	_____
nonane	_____	nonyne	_____	_____	$C_8H_{17}CHO$	_____
_____	$C_{10}H_{22}$	_____	_____	_____	_____	$C_{10}H_{20}O$

8. Name the following compounds.

(1)
$$CH_3$$
$$CH_3-\overset{\overset{\displaystyle CH_3}{|}}{C}=CH_2$$
_____

(2) $CH_3-CH_2-CH_2-C{\equiv}CH$ _____

(3)
$$CH_3-CH=\overset{\overset{\displaystyle CH_3}{|}}{C}-\overset{\overset{\displaystyle CH_3}{|}}{CH}-CH_3$$
_____

(4)
$$CH_3-\overset{\overset{\displaystyle CH_3}{|}}{CH}-CH_2-CH_2-CH=CH_2$$
_____

(5)
$$CH_3-CH_2-CH=\overset{\underset{\displaystyle CH_3}{|}}{C}-CH_3$$
_____

(6)
$$CH_3-\overset{\underset{\displaystyle CH_3}{|}}{CH}-\overset{\underset{\displaystyle CH_3}{|}}{C}=CH_2$$
_____

9. Name the aromatic compounds, using an appropriate system. You might choose to use either the numbering system or the o, m, p system if there are two substitutents on the benzene ring.

(1) 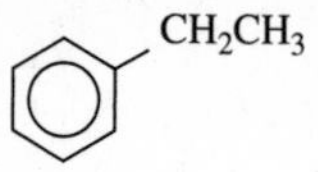 — CH_2CH_3 on benzene ring

_____

(2) 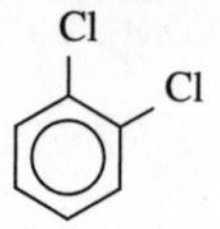 — Cl, Cl on benzene ring

_____

(3) — CH_3 (top), CH_3 (bottom) on benzene ring

_____

(4) 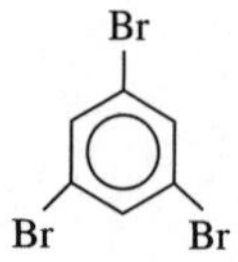 — Br, Br, Br on benzene ring

_____

(5) 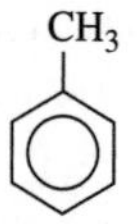 — CH_3 on benzene ring

_____

(6) 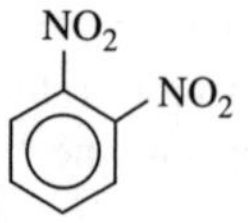 — NO_2, NO_2 on benzene ring

_____

10. Draw the structures for the following compounds.

(1) bromobenzene   _____

(2) ethylbenzene   _____

(3) phenol   _____

(4) p-bromoaniline   _____

(5) o-chloronitrobenzene   _____

(6) 1,3,5-trinitrobenzene   _____

11. Draw the structural formulas for the following alcohols.

(1) 2-methyl-2-butanol

(2) 3-chloro-l-butanol

(3) 2,2-dimethyl-l-propanol

(4) 3-methyl-2-butanol

(5) 2,3-dimethyl-l-pentanol

12. Identify the following alcohols as primary, secondary, tertiary, or polyhydroxy.

(1)
$$CH_3 - CH - \overset{H}{\underset{OH}{C}} - CH_3$$
$$\quad\quad\quad CH_3$$
_____

(2)
$$CH_3 - CH_2 - \overset{OH}{CH} - CH_3$$
_____

(3) $HO-CH_2-CH_2-OH$ _____

(4) $CH_3CH_2CH_2CH_2OH$ _____

$$CH_3$$
$$|$$
(5) 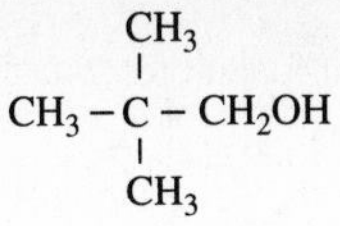 $CH_3 - C - CH_2OH$ _____
$$|$$
$$CH_3$$

(6) 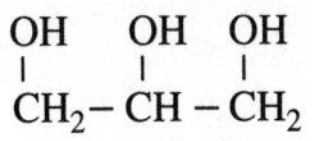
$$OH \quad OH \quad OH$$
$$| \qquad | \qquad |$$
$$CH_2 - CH - CH_2$$ _____

(7) $CH_3 - OH$ _____

(8) 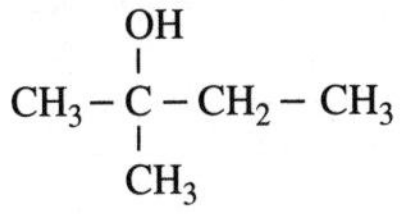
$$OH$$
$$|$$
$$CH_3 - C - CH_2 - CH_3$$ _____
$$|$$
$$CH_3$$

(9) 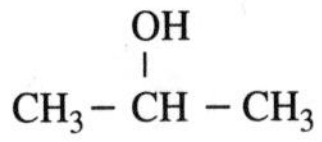
$$OH$$
$$|$$
$$CH_3 - CH - CH_3$$ _____

(10) 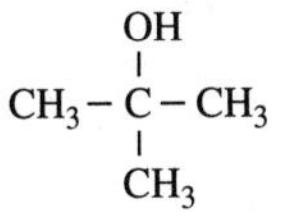
$$OH$$
$$|$$
$$CH_3 - C - CH_3$$ _____
$$|$$
$$CH_3$$

13.   Name the odd-numbered alcohols in question 12.

Write your answers below.

14.   Draw all possible isomers of the alcohol with the formula C_4H_9OH and name them.

Write your answers below.

15.   Fill in the blank space or circle the appropriate response.

Ethanol is one of our oldest and best known (1) <u>ketones/alcohols</u>. The preparation by the process known as (2) _____ is still carried out today. The raw materials are usually (3) _____ and (4) _____. A biological catalyst, called a(n) (5) <u>soap/enzyme</u>, is employed in natural fermentation to convert the raw materials into ethanol and carbon dioxide. As a drug, ethanol has been shown to be a (6) <u>depressant/stimulant</u>, which is contrary to many people's beliefs. For nonfood uses, ethanol is made unfit for drinking by a process called (7) _____. This process in effect poisons the ethanol.

16.   Give the names for the following compounds.

(1) $CH_3 - CH_2 - O - CH_2 - CH_3$

(2) $CH_3 - O - CH_2 - CH_3$

(3) 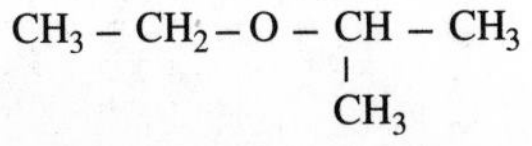
$$CH_3 - CH_2 - O - CH - CH_3$$
$$|$$
$$CH_3$$

17. Match the name of the functional group with the generalized formula.

(1) alkane   _____   a. RX

(2) aldehyde   _____   b. $R - \overset{\overset{\textstyle O}{\|}}{C} - R$

(3) alkyl halide   _____   c. R—H

(4) ether   _____   d. $R-CH=CH_2$

(5) ketone   _____   e. R—OH

(6) alcohol   _____   f. $R-C\equiv CH$

(7) alkyne   _____   g. R—O—R

(8) alkene   _____   h. R—CHO

18. After looking at the following list of structural formulas, match the names with the various formulas. Some of the names will be IUPAC and some will be common names.

(1) CH_3-O-CH_3 _____   a. ethanol

(2) $H_2C=O$ _____   b. formaldehyde

(3) CH_3CH_2OH _____

(4) $CH_3-\overset{\overset{\textstyle O}{\|}}{C}-CH_2-CH_3$ _____   c. t-butyl alcohol

(5) $CH_3-\overset{\overset{\textstyle OH}{|}}{CH}-CH_3$ _____   d. methoxymethane

(6) $HOCH_2CH_2OH$ _____   e. 1,2-ethanediol

(7) $CH_3(CH_2)_3CH_2OH$ _____   f. 3-methyl-2-butanol

(8) $CH_3-\overset{\overset{\textstyle CH_3}{|}}{\underset{\underset{\textstyle CH_3}{|}}{C}}-OH$ _____   g. butanal

(9) $\overset{\textstyle CH_3 \diagdown}{\underset{\textstyle CH_3 \diagup}{}}CH-\overset{\overset{\textstyle OH}{|}}{CH}-CH_3$ _____   h. 2-methylpropanal

(10) $CH_3(CH_2)_2CHO$ _____   i. 1-pentanol

(11) $CH_3-\overset{\overset{\textstyle }{}}{\underset{\underset{\textstyle CH_3}{|}}{CH}}-CHO$ _____   j. methyl ethyl ketone

   k. 2-propanol

19. From the list of formulas given for question 18, identify each one as an alcohol, ether, aldehyde, or ketone.

(1) _____   (2) _____

(3) _____   (4) _____

(5) _____   (6) _____

(7) _____ (8) _____

(9) _____ (10) _____

(11) _____

20. Name the organic acids using the IUPAC system.

(1) 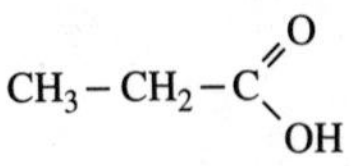 $CH_3-CH_2-C\overset{O}{\underset{OH}{}}$ _____

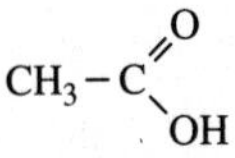

(2) $CH_3-C\overset{O}{\underset{OH}{}}$ _____

(3) $CH_3-CH_2-\overset{CH_3}{\underset{}{CH}}-C\overset{O}{\underset{OH}{}}$ _____

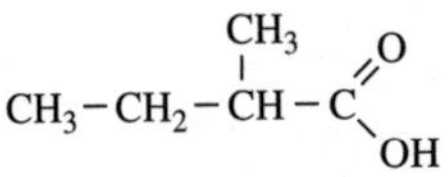

(4) _____

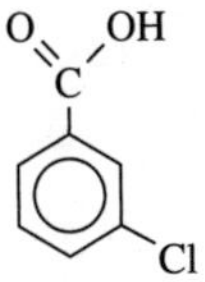

(5) 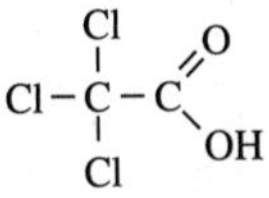 $Cl-\overset{Cl}{\underset{Cl}{C}}-C\overset{O}{\underset{OH}{}}$ _____

(6) _____

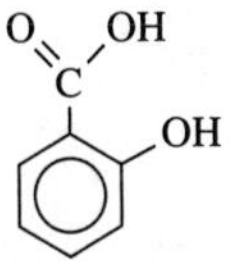

21. Fill in the blank spaces with the chemical formula or the name of the following organic acids.

Name	**Formula**
(1) formic acid	_____
(2) _____	CH_3COOH
(3) _____	CH_3CH_2COOH
(4) salicylic acid	_____
(5) benzoic acid	_____

22. Write the structural formulas for:

(1) Ethyl ethanoate

(2) Propyl methanoate

(3) Methyl benzoate

(4) Isopropyl ethanoate

23. Name the following esters, using the IUPAC system. Remember to name the acid portion of the ester last.

(1) 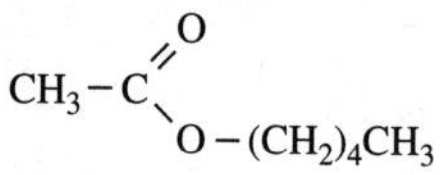 $CH_3-C\overset{O}{\underset{O-(CH_2)_4CH_3}{}}$ _____

(2) $CH_3-(CH_2)_2C\overset{O}{\underset{O-CH_2CH_3}{}}$ _____

(3) $HC\overset{O}{\underset{O-CH_2CH<\overset{CH_3}{CH_3}}{}}$ _____

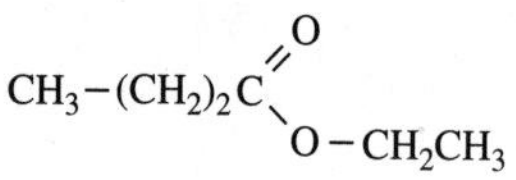

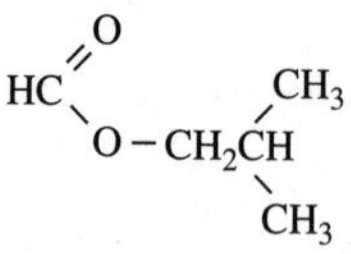

24. Match the monomer molecules listed in the left-hand column with the generalized formulas for the corresponding polymers in the right-hand column. There may be more than one monomer associated with a polymer, as in a copolymer.

(1) $CH_2=CCl_2$ _____

(2) 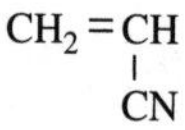 $\underset{CN}{CH_2=CH}$ _____

(3) $CH_2=\underset{}{\overset{CH_3}{C}}-C\overset{O}{\underset{OCH_3}{}}$ _____

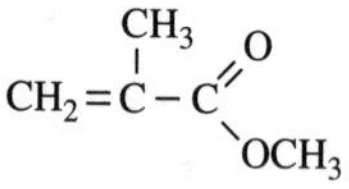

(4) 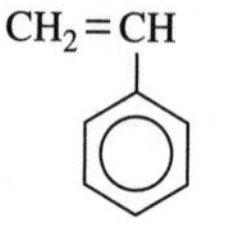 $CH_2=CH$ — ⬡ _____

(5) $CH_2=CF_2$ _____

(6) 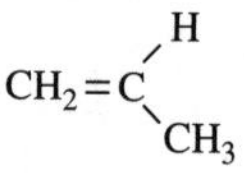 $CH_2=C\overset{H}{\underset{CH_3}{}}$ _____

a. $-(CH_2=CF_2)_n$

b. 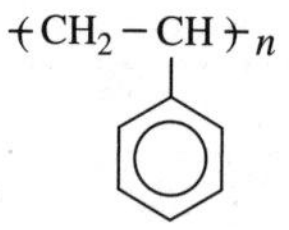 $\underset{⬡}{+CH_2-CH+_n}$

c. 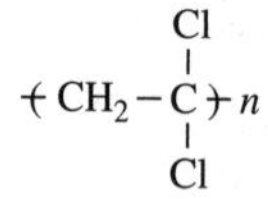 $+CH_2-\underset{Cl}{\overset{Cl}{C}}+_n$

d. 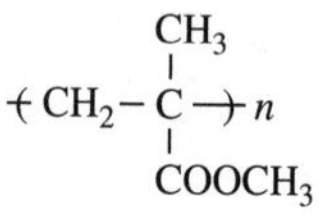 $+CH_2-\underset{COOCH_3}{\overset{CH_3}{C}}+_n$

e. 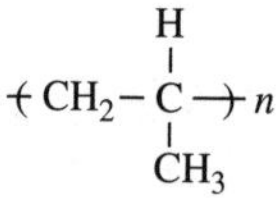 $+CH_2-\underset{CH_3}{\overset{H}{C}}+_n$

f. $+OCH_2CH_2O\ \overset{O}{\underset{}{C}}-⬡-\overset{O}{\underset{}{C}}+_n$

g. $+CH_2-\underset{CN}{CH}+_n$

h. $+CH_2-\underset{CH_3}{\overset{CH_3}{C}}+_n$

— 174 —

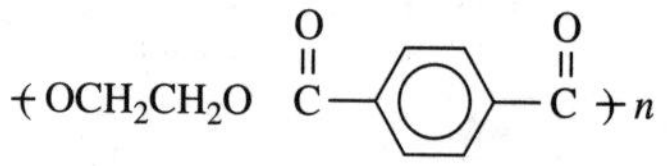

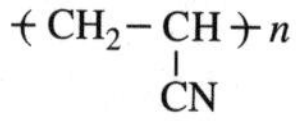

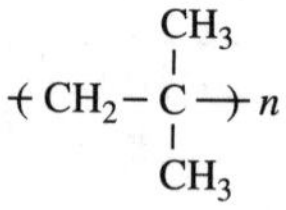

Challenge Problem

25. Choose the correct name of the ester formed from 3-methylbutanoic acid and *n*-hexanol.

    a) 3-methylbutyl *n*-hexanoate

    b) *n*-hexyl 3-methylpentanoate

    c) 2-methyl-4-decanoate

    d) *n*-hexyl 3-methylbutanoate

26. Which of the following names are incorrect? (Hint, you may want to draw the compound for those that are not already shown.)

    (1) 4-n-butyl-6-ethyl-5,5-dimethyl-4-octanol

    (2) 3,3-dimethyl-4-ethyl-4-propylbutanal

    (3)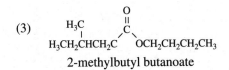
        2-methylbutyl butanoate

    (4) 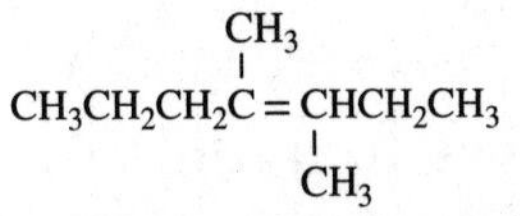 $CH_3CH_2CH_2C = CHCH_2CH_3$ with CH_3 above and CH_3 below
        3, 4-dimethyl-3-heptane

ANSWERS TO QUESTIONS AND PROBLEMS

1. (1) 4   (2) covalent   (3) tetrahedral   (4) itself
    (5) 2   (6) 4   (7) 6   (8) single
    (9) triple

2. Carbon tetrabromide

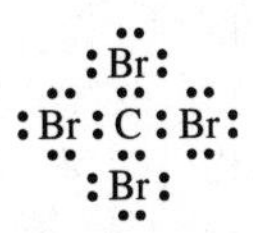

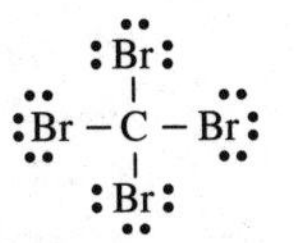

    Acetylene

    H : C ::: C : H   or   H—C≡C—H

    Methane

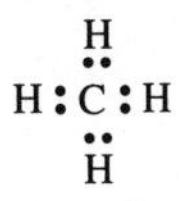

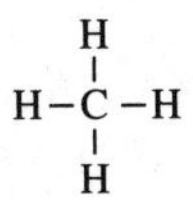

3. 2-iodubutane

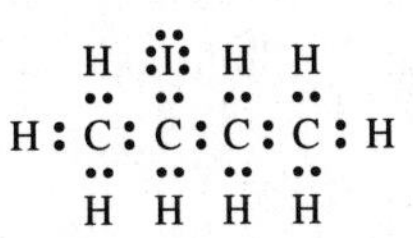

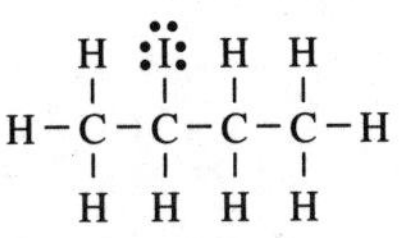

    pentane

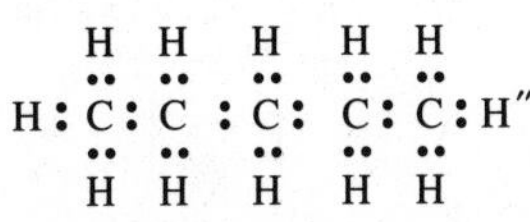

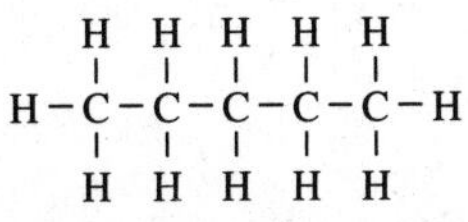

4. (1) n-heptane   (2) 3-methylhexane   (3) 3-ethylpentane
    (4) 2,2,3-trimethylbutane   (5) 3,3-dimethylpentane

5. (1) 2-methylbutane   (2) 1-chloro-3,4,4-trimethylhexane
    (3) 2-chlorobutane   (4) 2-bromo-1-chloropropane
    (5) 3-iodo-2-methylpentane   (6) 1-chloro-2-methylpropane
    (7) 2-fluoro-2-methylpropane   (8) 3,3-dimethylpentane

6. Four possible isomers of heptanol (where the carbon chain is unbranched).
There are seven carbons on heptane so all the possible structures would be as follows:

$$
\begin{array}{c}
\overset{6}{(OH)}\ \overset{5}{(OH)}\ \overset{4}{(OH)}\ \overset{3}{(OH)}\ \overset{2}{(OH)} \\
|\quad |\quad |\quad |\quad | \\
\overset{7}{(HO)} - C - C - C - C - C - C - C - \overset{1}{(OH)} \\
|\quad |\quad |\quad |\quad |\quad |\quad |
\end{array}
$$

The parentheses around the symbol for OH, are used to indicate all the possible locations to place a hydroxyl group on the heptane chain. There are seven structures to look at for duplicates. Numbers 1 and 7 are identical since the molecule is the same when looked at from either end. Numbers 2 and 6 are identical, as are 3 and 5. Number 4 is unique. Therefore there are four possible isomers of the monohydroxide of heptane (heptanol).

$CH_3-CH_2-CH_2-CH_2-CH_2-CH_2-CH_2OH$

$CH_3-CH_2-CH_2-CH_2-CH_2-CHOH-CH_3$

$CH_3-CH_2-CH_2-CH_2-CHOH-CH_2-CH_3$

$CH_3-CH_2-CH_2-CHOH-CH_2-CH_2-CH_3$

7. Members that were already in the table are shown in italics.

Alkanes		Alkynes		Aldehydes		
methane	CH_4	*Must have $\geq$ 2 C's*		methanal*	*HCHO*	*CH₂O*
ethane	C_2H_6	*ethyne*	C_2H_2	ethanal**	*CH₃CHO*	*C₂H₄O*
propane	C_3H_8	*propyne*	C_3H_4	*propanal*	C_2H_5CHO	C_3H_6O
butane	C_4H_{10}	butyne	C_4H_6	*butanal*	C_3H_7CHO	C_4H_8O
pentane	C_5H_{12}	pentyne	C_5H_8	*pentanal*	C_4H_9CHO	$C_5H_{10}O$
hexane	C_6H_{14}	*hexyne*	C_6H_{10}	hexanal	$C_5H_{11}CHO$	$C_6H_{12}O$
heptane	C_7H_{16}	heptyne	C_7H_{12}	heptanal	$C_6H_{13}CHO$	$C_7H_{14}O$
octane	C_8H_{18}	octyne	C_8H_{14}	*octanal*	$C_7H_{15}CHO$	$C_8H_{16}O$
nonane	C_9H_{20}	*nonyne*	C_9H_{16}	nonanal	$C_8H_{17}CHO$	$C_9H_{18}O$
decane	$C_{10}H_{22}$	decyne	$C_{10}H_{18}$	decanal	$C_9H_{19}CHO$	$C_{10}H_{20}O$

*methanal is more commonly known as formaldehyde
**ethanal is more commonly known as acetaldehyde

Notice the trend in each of the series: each successive member has a "CH₂" added to it regardless of series. E.g., "meth" to "eth" differ by one C and two H's, whether is it methane to ethane, or methanal to ethanal.

8. (1) 2-methylpropene   (2) 1-pentyne   (3) 3,4-dimethyl-2-pentene
    (4) 5-methyl-1-hexene   (5) 2-methyl-2-pentene   (6) 2,3-dimethyl-1-butene

9. (1) ethylbenzene   (2) ortho-dichlorobenzene or 1,2-dichlorobenzene
    (3) para-dimethylbenzene or 1,4-dimethylbenzene   (4) 1,3,5-tribromobenzene
    (5) toluene or methylbenzene   (6) ortho-dinitrobenzene or 1,2-dinitrobenzene

10. (1) Br [benzene ring]   (2) CH₂CH₃ [benzene ring]   (3) OH [benzene ring]

    (4) NH₂ [benzene ring] Br   (5) [benzene ring] Cl NO₂   (6) NO₂ [benzene ring] NO₂ NO₂

11.   (1)

$$CH_3CH_2 \overset{\overset{\displaystyle CH_3}{|}}{\underset{\underset{\displaystyle OH}{|}}{C}} - CH_3$$

    (2)

$$CH_3 \overset{\overset{\displaystyle Cl}{|}}{C}HCH_2CH_2OH$$

    (3)

$$CH_3 - \overset{\overset{\displaystyle CH_3}{|}}{\underset{\underset{\displaystyle CH_3}{|}}{C}}CH_2OH$$

  (4)

$$CH_3 - \overset{\overset{\displaystyle CH_3}{|}}{C}H - \overset{\overset{\displaystyle OH}{|}}{C}H - CH_3$$

    (5)

$$CH_3 - CH_2\overset{\overset{\displaystyle CH_3}{|}}{C}H - \overset{\overset{\displaystyle CH_3}{|}}{C}H - CH_2OH$$

12.   (1)   secondary   (2)   secondary   (3)   polyhydroxy
     (4)   primary   (5)   primary   (6)   polyhydroxy
     (7)   primary   (8)   tertiary   (9)   secondary
  (10)   tertiary

13.   (1)   3-methyl-2-butanol   (2)   1,2-ethanediol   (3)   2,2-dimethylpropanol
    (4)   methanol   (5)   2-propanol (isopropanol)

14.   There are four possible isomers with a formula of C_4H_9OH:

  (1) $CH_3CH_2CH_2CH_2OH$
      1-butanol

     (2) $CH_3CH_2\overset{\overset{\displaystyle OH}{|}}{C}HCH_3$
         2-butanol

  (3) $CH_3 - \overset{\overset{\displaystyle OH}{|}}{\underset{\underset{\displaystyle CH_3}{|}}{C}} - CH_3$
      2-methyl-2-propanol

     (4) $CH_3 - \overset{\overset{\displaystyle }{}}{C}H - CH_2OH$
             $|$
           CH_3
        2-mythyl-1-propanol

15.   (1)   alcohols   (2)   fermentation   (3)   starch
    (4)   sugar   (5)   enzyme   (6)   depressant
    (7)   denaturing

16.   (1)   diethyl ether
    (2)   methyl ethyl ether
    (3)   ethyl isopropyl ether

17.   (1)   c   (2)   h   (3)   a   (4)   g
    (5)   b   (6)   e   (7)   f   (8)   d

18.   (1)   d   (7)   i
    (2)   b   (8)   c
    (3)   a   (9)   f
    (4)   j   (10)   g
    (5)   k   (11)   h
    (6)   e

19.   (1)   ether   (7)   alcohol
    (2)   aldehyde   (8)   alcohol
    (3)   alcohol   (9)   alcohol
    (4)   ketone   (10)   aldehyde
    (5)   alcohol   (11)   aldehyde
    (6)   alcohol (diol)

20.   (1)   propanoic acid   (2)   ethanoic acid or acetic acid
    (3)   2-methylbutanoic acid   (4)   3-chlorobenzoic acid
    (5)   trichloroethanoic acid   (6)   2-hydroxybenzoic acid

21. (1) HCOOH (2) acetic acid

(3) propionic or propanoic acid (4)

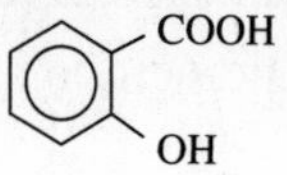

(5)

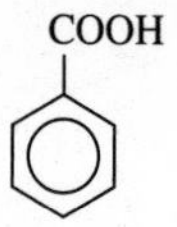

22. (1) 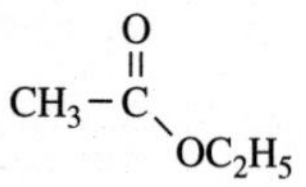   (2) 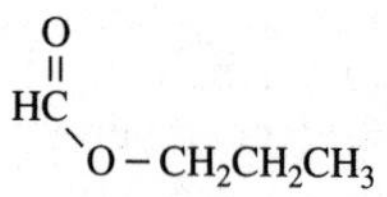   (3)

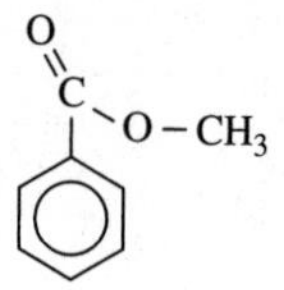

(4)

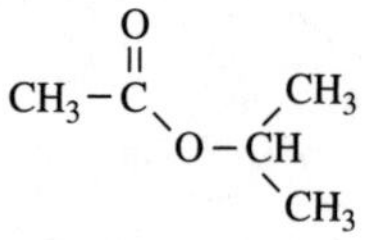

23. (1) pentyl ethanoate (2) ethyl butanoate (3) isobutyl methanoate

24. (1) c (2) g (3) d
    (4) b (5) a (6) e

25. d

26. (1)

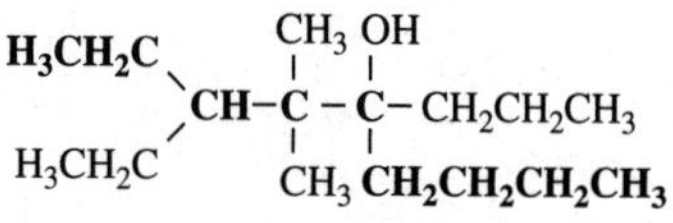

3-ethyl-4,4-dimethyl-5-n-propyl-5-nonanol
(the correct main chain is in bold)

(2)

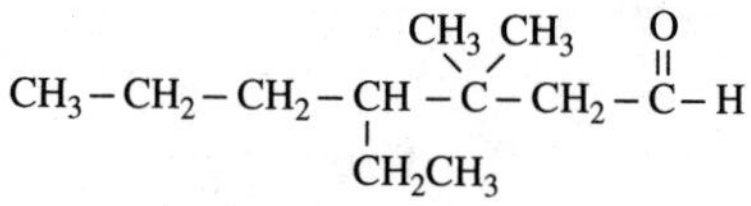

4-ethyl-3,3-dimethylheptanal

(3)

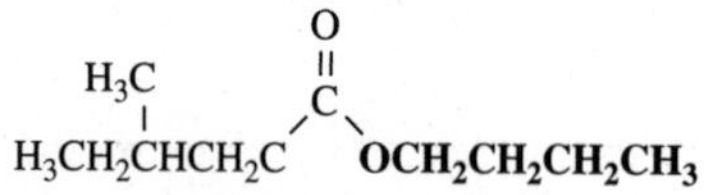

butyl 4-methylpentanoate
(the alcohol portion is in bold and is named first)

(4) This name is correct.

CHAPTER TWENTY

Introduction to Biochemistry

SELECTED CONCEPTS REVISITED

A very brief summary of some biochemical compounds covered in this chapter are:

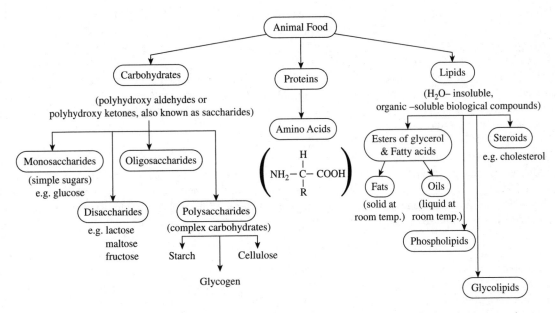

Nucleic acids, DNA and RNA, are polymers of nucleotides. A nucleotide consists of phosphate-sugar-base.

DNA: sugar is *deoxyribose*	bases are A, G, C, *T*	*double*-stranded
RNA: sugar is **ribose**	bases are A, G, C, **U**	**single**-stranded

COMMON PITFALLS TO AVOID

An open chain sugar has an aldehyde with which an alcohol reacts to close the chain. To find the original carbon of the aldehyde in a closed ring sugar, look for the carbon having two oxygens directly attached to the carbon.

Peptide chains are named from N-terminal to C-terminal. Be sure to look for the free NH_2 or NH_3^+ group at the end of the chain. Remember that the sequence to look for to verify that you are following the peptide chain (actual atoms to follow are in bold) is - "**N**" - "α**C**—**H**" - "**C**=**O**" - "**N**" - "α**C**—**H**" - "**C**=**O**" - (nitrogen, alpha carbon, carbonyl). If you find something else, you have ventured down a side chain.

RECAP SECTION

Chapter 20 is a survey chapter covering the topic of biochemistry in much the same manner that Chapter 19 surveys the broad field of organic chemistry. Four major classes largely cover the huge quantities of biomolecules: carbohydrates, lipids, proteins, and nucleic acids. Carbohydrates, often with an empirical formula of $(CH_2O)_n$, may serve as energy sources or fuel molecules. Lipids, whether fats or oils, and proteins, built from amino acids, can have these functions also, but also serve important roles such as hormones and catalysts, among others. The general characteristics of lipids and the chemical composition and functions of amino acids and proteins are all covered in this chapter. Nucleic acids and proteins in the form of enzymes serve functional roles and may act as information molecules. With DNA being a common term, even in the media among nonscientists, its general chemical structure and its functions in genetics would be worthwhile knowledge for the well-informed individual. If you are anticipating a career in an allied health field, this chapter should be of particular interest and it would be wise to master the content, do further reading, and take additional courses in this area.

SELF-EVALUATION SECTION

1. Fill in the blank space or circle the appropriate response.

The branch of chemistry that deals with the chemical reactions of living organisms and is called (1) _____ is one of the fastest moving and most exciting fields of science today. It wasn't very long ago that biochemists worked out the pathway by which carbon dioxide is chemically reduced by energy derived from the sun.

The entire process of trapping the radiant energy of the sun in chemical bond energy is called (2) _____. The products of this process are (3) _____ molecules, which can be single units named (4) _____, two single units bonded together named (5) _____, or long chain molecules composed of many units called (6) _____. The carbohydrates which are produced as a result of photosynthesis can be used for many purposes in an organism. They are a source of immediate energy or they may be stored in the form of starch. Green plants possess a rigid structure because of the polysaccharide, (7) _____, which indicates that carbohydrates can play a structural role in addition to an energy source role.

Two other important classes of food stuffs are (8) _____ and (9) _____. The oil seeds such as peanuts, soybeans, and cottonseed are important sources of oils and fats. An important commercial product, (10) _____, is made much the same way pioneers used to make it with lye, beef tallow, and a big iron kettle. The other energy compounds, the proteins, serve additional roles. Hair, fingernails, and connective tissue are structural proteins.

The biological catalysts, (11) _____, are proteins which serve a specialized vital role in metabolism. And, of course, the vegetable and animal protein we eat serves as food material. One of the important processes of digestion in the human stomach is to break down proteins into small sub-units. The single sub-units of proteins are called (12) _____, of which 20 are commonly found in all proteins. Amino acids join together to form proteins through (13) _____ linkages, which produces molecules of great diversity. The large number of different protein molecules do not just develop by chance. It is directed by yet another group of compounds whose structure and function has recently been determined. These compounds are (14) _____, which we abbreviate as (15) _____ and (16) _____. (17) _____ is found in the nucleus of every living cell, whereas most of the (18) _____ is found in the cell matrix, where it is involved in protein synthesis.

2. Match the name of a sugar with its structural formula, either open-chain or cyclic

(1) glucose   (4) fructose   (7) lactose
(2) maltose   (5) ribose
(3) galactose   (6) sucrose

(a) CH$_2$OH

O   OH

H   H   H   H

OH   OH

(b) H$-$C$=$O

H$-$C$-$OH

HO$-$C$-$H

HO$-$C$-$H

H$-$C$-$OH

CH$_2$OH

(c)   CH$_2$OH

H   O   H

H

HO   OH   H   OH

H   OH

(d)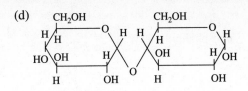

(e)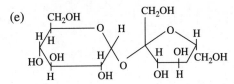

(f)

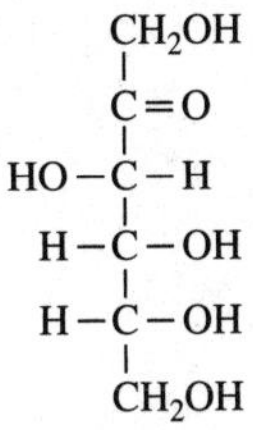

```
        CH₂OH
         |
         C=O
         |
  HO—C—H
         |
   H—C—OH
         |
   H—C—OH
         |
        CH₂OH
```

(g)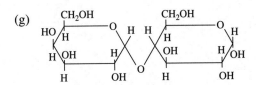

3. Classify the following carbohydrates as monosaccharides, disaccharides, or polysaecharides.

   (1) galactose   (6) erythrose
   (2) starch   (7) glycogen
   (3) glucose   (8) maltose
   (4) fructose   (9) ribose
   (5) lactose   (10) sucrose

4. List the monosaccharides in the following carbohydrates:

   (1) maltose   (2) lactose   (3) sucrose   (4) starch

5. Match each of the following phrases as being most descriptive of glucose, galactose, fructose or ribose.

   _____ (1) also known as levulose

   _____ (2) requires no digestion by humans

   _____ (3) an aldopentose

   _____ (4) sweetest of all the sugars; found in honey

   _____ (5) less than half as sweet as glucose

   _____ (6) an important component of RNA

   _____ (7) also known as blood sugar

   _____ (8) a ketohexose

   _____ (9) found in the urine of diabetics

   _____ (10) infants unable to metabolize this sugar may experience vomiting, diarrhea, or even death if not diagnosed early.

6. Match the name of a fatty acid with its structural formula.

   (1) Palmitoleic acid   (a) $CH_3(CH_2)_4CH=CHCH_2CH=CH(CH_2)_7COOH$
   (2) Linoleic acid   (b) $CH_3(CH_2)_7CH=CH(CH_2)_7COOH$
   (3) Oleic acid   (c) $CH_3(CH_2)_5CH=CH(CH_2)_7COOH$

7. Match the name of a lipid molecule with its structural formula.

(1) triacylglycerol (2) essential fatty acid (3) soap
(4) glycolipid   (5) steroid   (6) phospholipid

(a)

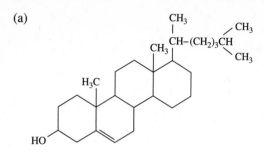

(b) 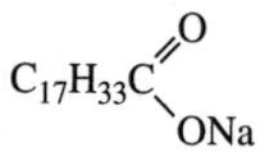 $C_{17}H_{33}C$ 〈^{O} ONa〉

(c)
$CH_2-O-\overset{O}{\overset{\|}{C}}-C_{17}H_{35}$

$CH-O-\overset{O}{\overset{\|}{C}}-C_{15}H_{31}$

$CH_2-O-\overset{O}{\overset{\|}{C}}-C_{11}H_{23}$

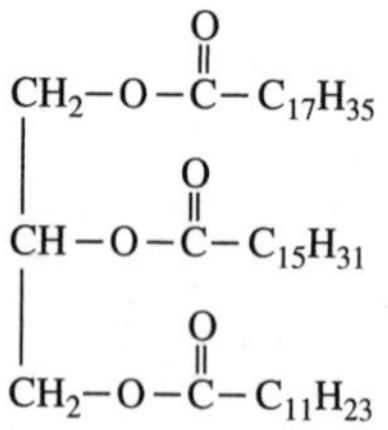

(d)
$CH_2-O-\overset{O}{\overset{\|}{C}}-R$

$CH-O-\overset{O}{\overset{\|}{C}}-R'$

$CH_2-O-\overset{O}{\overset{\|}{P}}-O-CH_2CH_2\overset{\oplus}{N}\overset{CH_3}{\underset{CH_3}{-}}CH_3$
                    $\underset{OH}{|}$

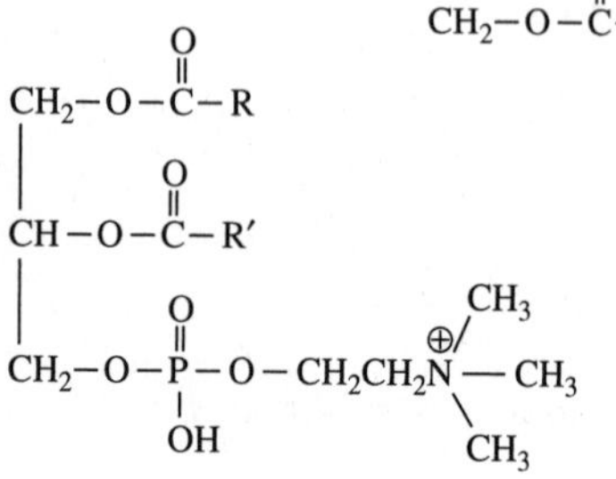

(e)
$CH_3(CH_2)_{12}-CH=CH-\overset{OH}{\underset{|}{CH}}-CH-\overset{O}{\overset{\|}{NHC}}-R$
                              $\underset{CH_2}{|}$

CH_2OH

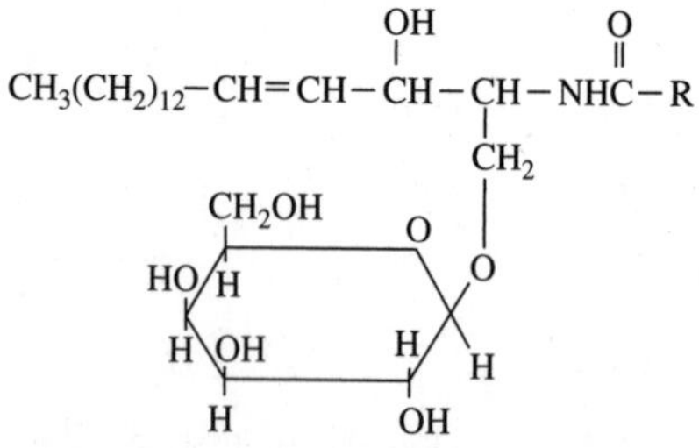

(f) $CH_3(CH_2)_4CH=CHCH_2CH=CH(CH_2)_7COOH$

8. Circle the appropriate response or fill in the blank spaces.

Lipids are a group of (1) <u>organic/inorganic</u> compounds classified on the basis of their solubility (i.e., they are soluble) in fat solvents such as (2) <u>ether, benzene, etc/water, salt solution, etc</u>. The three types of simple lipids are (3) _____, (4) _____, and (5) _____.

9. Fill in the blank spaces.

general structure of a
(1) _____ or a (2) _____

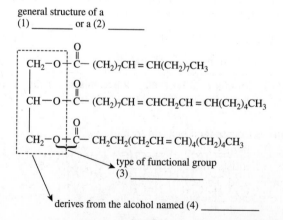

type of functional group
(3) _____

derives from the alcohol named (4) _____

— 182 —

Name the three fatty acids that are part of this structure:

(4) _____

(5) _____

(6) _____

Which of the three fatty acids above is the most unsaturated?

(7) _____

10. Assign the following as relevant to a fat or an oil.

(1) solid at room temperature _____

(2) mostly from vegetable sources _____

(3) high degree of unsaturation in fatty acid groups _____

(4) mostly from animal sources _____

(5) liquid at room temperature _____

(6) often fully saturated (no double bonds in the fatty acid groups) _____

11. From the list below, circle the foods high in protein content.

bread   melon   spaghetti   apples

fish   chichen   cheese   potatoes

carrots   nuts   eggs   beans

12. Fill in the blank space with either the name of the amino acid or its abbreviation.

Name	Abbreviation
(1) Glutamic acid	_____
(2) _____	val
(3) _____	ser
(4) Leucine	_____
(5) Cysteine	_____
(6) Histidine	_____
(7) _____	arg
(8) _____	pro
(9) Methionine	_____
(10) Phenylalanine	_____

13. Fill in the blank space or circle the appropriate response.

Proteins, which are polymers composed of (1) _____, serve two essential functions: that of (2) a source of oxygen/a structural role and (3) as enzymes/as activators. When proteins are hydrolyzed into their component amino acids, scientists find approximately (4) 30/20 usual amino acids with (5) two/three identical functional groups per amino acid. These two groups are (6) _____ and (7) _____. On a list of common amino acids there are eight that are classified as essential for (8) beef/humans. This means that these amino acids cannot be (9) _____ by an individual and must be supplied in the (10) _____. In addition to serving a variety of structural roles in an organism, proteins are also found to be biological catalysts or (11) _____.

14. Write the two possible dipeptide structures formed when alanine is joined to valine. For each structure, point out the peptide linkage. Name each dipeptide.

Write your structures here.

15. Fill in the blank space or circle the correct response.

The great variety of proteins found in any organism does not arise by chance. The substances that direct the course of protein synthesis are known as (1) _____. The two major types are known by abbreviations (2) _____ and (3) _____. There are several major differences between these types of nucleic acids. RNA contains the sugar (4) _____, and DNA contains (5) _____. DNA is a double-stranded (6) _____, and RNA is only (7) _____ stranded. In addition, one of the RNA bases is different. DNA contains the base (8) _____ and RNA contains (9) _____. In nucleic acids, the actual building blocks are phosphate esters. These molecules are called (10) _____. In the double-stranded helix of DNA, it has been found that (11) complementary/dissimilar pairing occurs between purine and pyrimidine bases. The genetic information contained in the cell's genes is found to be the exact sequence of nucleotides along the DNA strands. When the DNA replicates itself during normal cell division called (12) _____, the two daughter cells have the same double-stranded DNA sequences as the original cell. In human reproduction, cell splitting occurs by a different process called (13) _____, with the male and female both contributing to the new cell. When fertilization takes place the zygote has a normal complement of genes, half from the mother and half from the father. Each gene, which directs the synthesis of one protein, contains what is termed the genetic code. Each codon, for a specific amino acid, is composed of (14) _____ nucleotides in sequence. If, for some reason, the DNA structure is changed or a breakdown occurs in the protein synthesis process, a (15) _____ may occur. Most severe cases result in an organism's death, but sometimes the organism survives. For example, sickle cell anemia results from (16) two/one change(s) in an amino acid on one chain of hemoglobin.

Challenge Problems

16. Name the following pentapeptides.
    (1) ala-asp-arg-ala-asn
    (2) cys-ile-pro-ser-gly

17. An nonapeptide was known to contain three serine residues and two methionine residues. The following peptides were obtained after partial hydrolysis. Determine the sequence.

    met-ser-trp, ser-ser, met-leu-ser, trp-lys, ser-ile-met

ANSWERS TO QUESTIONS

1. (1) biochemistry   (2) photosynthesis   (3) carbohydrate
    (4) monosaccharides   (5) disaccharides   (6) polysaccharides
    (7) cellulose   (8) lipids   (9) proteins
    (10) soap   (11) enzymes   (12) amino acids
    (13) peptide   (14) nucleic acids   (15) DNA
    (16) RNA   (17) DNA   (18) RNA

2. (1) –(c),   (2) –(d),   (3) –(b),   (4) –(f),   (5) –(a)   (6) –(e),   (7) –(g)

3. (1) mono   (2) poly   (3) mono
    (4) mono   (5) di   (6) mono
    (7) poly   (8) di   (9) mono
    (10) di

4. (1) maltose is 2 glucose units
    (2) lactose is galactose plus glucose
    (3) sucrose is glucose plus fructose
    (4) starch is glucose units

5. (1) fructose (2) glucose (3) ribose   (4) fructose (5) galactose
    (6) ribose   (7) glucose (8) fructose (9) glucose   (10) galactose

6. (1) (c)   (2) (b)   (3) −(a)

7. (1) −(c),   (2) −(f),   (3) −(b),
    (4) −(e),   (5) −(a),   (6) −(d)

8. (1) organic   (2) ether, benzene, etc   (3), (4), (5) fats, oils, waxes

9. (1), (2) fat, oil   (3) ester   (4) glycerol
    (4) oleic acid   (5) linoleic acid   (6) arachidonic acid   (7) arachidonic acid

10. (1) fat   (2) oil   (3) oil   (4) fat   (5) oil   (6) fat

11. Fish, chicken, nuts, cheese, eggs, beans.

12. (1) glu   (2) Valine   (3) Serine
    (4) leu   (5) cys   (6) his
    (7) Arginine   (8) Proline   (9) met
    (10) phe

13. (1) amino acids   (2) a structural role (3) as enzymes
    (4) 20   (5) two   (6) amino
    (7) carboxyl   (8) humans   (9) synthesized
    (10) diet or food (11) enzymes

14.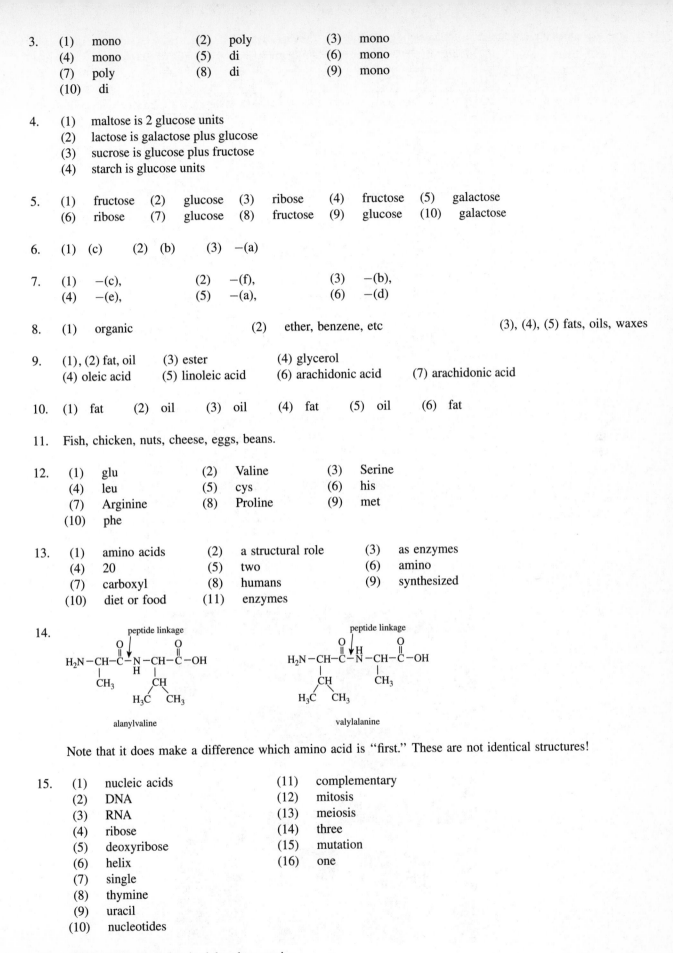

alanylvaline

valylalanine

Note that it does make a difference which amino acid is "first." These are not identical structures!

15. (1) nucleic acids   (11) complementary
    (2) DNA   (12) mitosis
    (3) RNA   (13) meiosis
    (4) ribose   (14) three
    (5) deoxyribose   (15) mutation
    (6) helix   (16) one
    (7) single
    (8) thymine
    (9) uracil
    (10) nucleotides

16. (1) alanylaspartylarginylalanylasparagine
    (2) cystylisoleucylprolylserylglycine

17. The nonapeptide would be: met-leu-ser-ser-ile-met-ser-trp-lys